Mathematics Workbook

10-Minute Maths Tests

AGE 8–10

David E Hanson

About the author

David Hanson has over 40 years' experience of teaching and has been leader of the Independent Schools Examinations Board (ISEB) 11+ Maths setting team, a member of the ISEB 13+ Maths setting team and a member of the ISEB Editorial Endorsement Committee.

Every effort has been made to trace all copyright holders, but if any have been inadvertently overlooked the publishers will be pleased to make the necessary arrangements at the first opportunity.

Although every effort has been made to ensure that website addresses are correct at time of going to press, Galore Park cannot be held responsible for the content of any website mentioned in this book. It is sometimes possible to find a relocated web page by typing in the address of the home page for a website in the URL window of your browser.

Hachette UK's policy is to use papers that are natural, renewable and recyclable products and made from wood grown in sustainable forests. The logging and manufacturing processes are expected to conform to the environmental regulations of the country of origin.

Orders: please contact Bookpoint Ltd, 130 Milton Park, Abingdon, Oxon OX14 4SB. Telephone: +44 (0)1235 827827. Lines are open 9.00a.m.–5.00p.m., Monday to Saturday, with a 24-hour message answering service. Visit our website at www.galorepark.co.uk for details of other revision guides for Common Entrance, examination papers and Galore Park publications.

Published by Galore Park Publishing Ltd
An Hachette UK company
338 Euston Road, London, NW1 3BH
www.galorepark.co.uk

Text copyright © David Hanson 2014
The right of David Hanson to be identified as the author of this Work has been asserted by him in accordance with sections 77 and 78 of the Copyright, Designs and Patents Act 1988

Impression number 10 9 8 7 6 5 4 3 2 1
2018 2017 2016 2015 2014

All rights reserved. No part of this publication may be sold, reproduced, stored in a retrieval system, or transmitted, in any form or by any means, electronic, mechanical, photocopying, recording, or otherwise, without either the prior written permission of the copyright owner or a licence permitting restricted copying issued by the Copyright Licensing Agency, Saffron House, 6–10 Kirby Street, London EC1N 8TS.

Typeset in India
Printed in Italy
Illustrations by Aptara, Inc.

A catalogue record for this title is available from the British Library.

ISBN: 978 1 471 829 61 1

Contents and test results

	Page	Completed	Marks	%	Time taken
Number test 1	7				
Number test 2	8				
Number test 3	9				
Number test 4	10				
Number test 5	11				
Number test 6	12				
Number test 7	13				
Number test 8	14				
Number test 9	15				
Number test 10	16				
Calculations test 1	17				
Calculations test 2	18				
Calculations test 3	19				
Calculations test 4	20				
Calculations test 5	21				
Calculations test 6	22				
Calculations test 7	23				
Calculations test 8	24				
Calculations test 9	25				
Calculations test 10	26				
Problem solving test 1	27				
Problem solving test 2	28				
Problem solving test 3	29				
Problem solving test 4	30				
Problem solving test 5	31				
Problem solving test 6	32				
Problem solving test 7	33				
Problem solving test 8	34				
Problem solving test 9	35				
Problem solving test 10	36				
Pre-algebra test 1	37				
Pre-algebra test 2	38				
Pre-algebra test 3	39				
Pre-algebra test 4	40				

	Page	Completed	Marks	%	Time taken
Pre-algebra test 5	41				
Pre-algebra test 6	42				
Pre-algebra test 7	43				
Pre-algebra test 8	44				
Pre-algebra test 9	45				
Pre-algebra test 10	46				
Shape, space and measures test 1	47				
Shape, space and measures test 2	48				
Shape, space and measures test 3	49				
Shape, space and measures test 4	50				
Shape, space and measures test 5	51				
Shape, space and measures test 6	52				
Shape, space and measures test 7	53				
Shape, space and measures test 8	54				
Shape, space and measures test 9	55				
Shape, space and measures test 10	56				
Handling data test 1	57				
Handling data test 2	58				
Handling data test 3	59				
Handling data test 4	60				
Handling data test 5	61				
Handling data test 6	62				
Handling data test 7	63				
Handling data test 8	64				
Handling data test 9	65				
Handling data test 10	66				
Mixed test 1	67				
Mixed test 2	68				
Mixed test 3	69				
Mixed test 4	70				
Mixed test 5	71				
Mixed test 6	72				
Mixed test 7	73				
Mixed test 8	74				
Mixed test 9	75				
Mixed test 10	76				

Answers to all the questions in this book can be found in the pull-out section in the middle.

Introduction

The material in this workbook takes account of recent developments in the National Curriculum but, for convenience, the earlier 'strand' divisions are used as a basis for grouping the topics.

The book consists of 70 tests. There are 10 tests in each of the six subject 'strands' and 10 mixed tests.

Strands in this book	Topics covered	New National Curriculum
Number	Properties of numbers: multiples and factors, counting backwards and forwards, sequences, prime numbers, negative numbers, squares and cubes	Number (1) – number and place value
	Place value and ordering: words and numerals, Roman numerals, place value including abacus, multiplying by powers of 10	with some elements of Number (4)
	Estimation and approximation: estimating position on a number line, rounding to the nearest power of 10, nearest integer and 1 decimal place	
	Fractions, decimals and percentages: equivalent fractions, ordering fractions, adding and subtracting fractions, multiplying and dividing fractions, equivalence of fractions, decimals and percentages, fractions and percentages of quantities	Number (4) – fractions (including decimals and percentages)
Calculations	Number operations: understanding and using basic number facts, order of operations including use of brackets	Number (2) – addition and subtraction
	Mental strategies: practice using a variety of strategies, including making use of known facts	Number (3) – multiplication and division
	Written methods: practice in all methods with integers and decimals	
	Interpreting and checking results: practice using even and odd number facts, and approximations	
Problem solving	Decision making and reasoning about numbers or shapes: looking for patterns, sequences	Number (1)
		Number (2)
	Real-life mathematics: applying mathematical skills to everyday problems	Number (3)
		Number (4)
Pre-algebra	Missing numbers in number sentences, simple word formulae, sequences, number machines	Number (2)
		Number (3)
Shape, space and measures	Measures: measurement units and conversions, reading scales, area and perimeter of rectilinear shapes, time	Measurement
	Shape: names and symmetry of plane shapes, polygons, names and nets of solid shapes	Geometry (1) – properties of shapes
	Space: names of angles, calculating angles in a right angle and at a straight line, equal sides and angles of plane shapes, points on a co-ordinate grid, translation of a shape on a co-ordinate grid	Geometry (2) – position and direction
Handling data	Data tables, pictograms, bar charts, tallies, frequency diagrams, simple Carroll diagrams and Venn diagrams, line graphs, likelihood	Statistics
Mixed tests	A selection of questions from all six of the strands above.	

The strand tests:

- consist of 10 questions printed in two columns on one page
- can be tackled in a suggested time of 10 minutes
- contain questions on different topics of the 'strand'
- cover the key ideas in the strand
- feature a gradual increase in difficulty through the ten tests of each strand
- follow the same general pattern, and are designed to:
  - build confidence
  - facilitate the identification of weak areas
  - provide practice in recalling facts and procedures
  - facilitate the monitoring of progress
  - encourage working quickly and accurately.

Two consecutive strand tests – for example 1 and 2, 3 and 4 or 5 and 6 – could be combined to form an assessment covering the whole strand, to be tackled in a time of 20 minutes.

The mixed tests consist of 10 questions from different sections of the strands.

Two consecutive mixed tests – for example 1 and 2, 3 and 4 or 5 and 6 – could be combined to form an assessment covering the whole curriculum, to be tackled in a time of 20 minutes.

Using the tests

The tests can be used in two main ways:

- Complete the test, as quickly as possible, recording the time taken.
- Do as much as possible in a fixed time.

Responses should be written or drawn on the page but additional paper should be available to do extra working if required.

Answers

Answers to the questions can be pulled out of the middle of the book.

Where appropriate, answers involving fractions should be given in their simplest form.

Marks

Each question has a total of two marks. The number of parts in a question varies and it is left to the marker to decide on allocation of the 2 marks. For a question with an odd number of parts, give half marks to the easier parts and whole marks to the more difficult parts. It is left to the discretion of markers to award a half mark if this is considered appropriate, particularly in the early stages or for a weak pupil.

Notes

In decimal abacus pictures, the horizontal line above the 4th bead position is an aid to rounding.

Looking at the spike immediately to the right of the 'cut-off', if there are beads above the line (5 or more beads on the spike), the number is rounded up; if there are no beads above the line (4 beads or fewer on the spike), the number is rounded down.

To the nearest ten, this number is 2550. To the nearest hundred it is 2500. To the nearest thousand it is 3000.

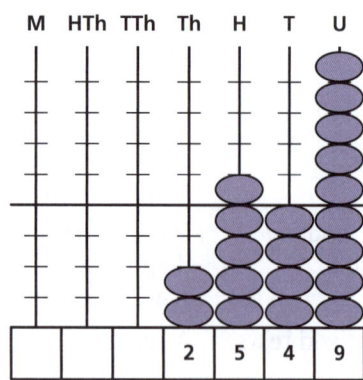

Number test 1

Final score
$\overline{20}$ ____ %

1 (a) On the number grid, shade all the multiples of 2 and circle all the multiples of 3

4	5	6	7	8	9
14	15	16	17	18	19
24	25	26	27	28	29
34	35	36	37	38	39

(b) Write down two numbers between 40 and 50 which are multiples of both 2 and 3

_____ _____

2 (a) What is the largest number which is a factor of both 12 and 18?

_____

(b) Complete the factor rainbow to show all of the factor pairs for 12

1   2 ____ ____ ____ 12

3 (a) Count forwards in 100s from 1000

1000 ____ ____ ____ ____

(b) Count backwards in 10s from 220

220 ____ ____ ____ ____

4 Write the next two terms in each sequence.

(a) 1, 5, 9, 13, _____, _____

(b) 1, 2, 4, 8, _____, _____

5 (a) Circle the prime numbers in the list below.

23   39   45   57   61   77   87

(b) List the prime numbers between 10 and 20

_____

(c) Write down two prime factors of 24

_____ _____

6 (a) What temperature is 4 degrees lower than 3 °C?

_____ °C

(b) What number is 7 more than ⁻3 (negative 3)?

_____

7 Write down

(a) the square of 4 _____

(b) 3^3 (the cube of 3) _____

(c) all of the square numbers between 10 and 30

_____

8 (a) Write **five hundred and twenty-four** in figures.

_____

(b) Write 3056 in words. _____

_____

(c) Write in figures the Roman number DCCLXXVI.

_____

9 (a) What is the value of the 8 in 508 075?

_____

(b) Draw beads on the abacus to represent the number which is 5 more than 508 075

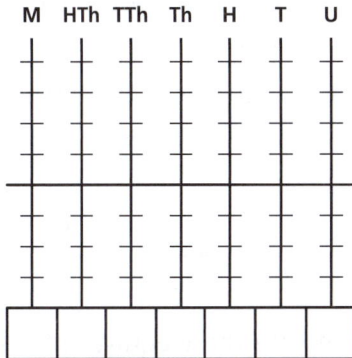

10 (a) Multiply 345 by 10 _____

(b) Divide 7080 by 100 _____

7

Number test 2

Final score

1 (a) If these numbers are arranged in order of size, which will be in the middle?

_____

345 534 354 543 453

(b) Write down the number which is exactly half way between 22 and 30 _____

2 (a) On the scale, mark the positions of 4 and ⁻2

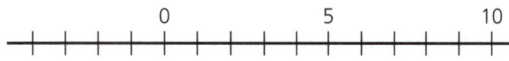

(b) On the scale, mark the positions of 0.2 and 1.3

3 (a) On the scale, mark the position of 5

(b) On the scale, mark the position of 25

4 (a) Round 7459 to the nearest 100

_____

(b) Round 909 090 to the nearest 10 000

_____

5 (a) Round 5.43 to the nearest whole number.

_____

(b) Round 17.56 to 1 decimal place.

_____

6 (a) Shade a quarter of this shape.

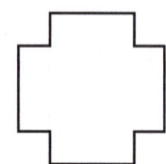

(b) Complete these equivalent fractions:

$\frac{1}{4} = \frac{}{8}$

(c) Write these fractions in order of increasing size: $\frac{1}{8}, \frac{1}{4}, \frac{1}{12}$

_____ _____ _____

(d) Write $1\frac{1}{2}$ as an improper fraction.

_____

7 (a) Add: $\frac{3}{5} + \frac{1}{5}$

_____

(b) Subtract: $\frac{6}{7} - \frac{4}{7}$

_____

8 (a) Multiply: $\frac{1}{2} \times 4$

_____

(b) Divide: $\frac{1}{4} \div 2$

_____

9 (a) Write the fraction $\frac{35}{100}$ as a decimal.

_____

(b) Write the decimal 0.43 as a fraction.

_____

(c) Write 49% as a fraction.

_____

(d) Write 65% as a decimal.

_____

10 What is

(a) 10% of 650 g _____ g

(b) $\frac{4}{5}$ of £100? £_____

Number test 3

Final score

20 ___ %

1 (a) On the number grid, shade all the multiples of 3 and circle all the multiples of 4

1	2	3	4	5	6
11	12	13	14	15	16
21	22	23	24	25	26
31	32	33	34	35	36

(b) Write down a number between 40 and 50 which is a multiple of both 3 and 4

_____

2 (a) What is the largest number which is a factor of both 45 and 30?

_____

(b) Complete the factor rainbow to show all of the factor pairs for 28

1 2 ____ ____ ____ 28

3 (a) Count forwards in 10s from 975

975 ____ ____ ____ ____

(b) Count backwards in 100s from 1040

1040 ____ ____ ____ ____

4 Write the next two terms in each sequence.

(a) 1, 4, 7, 10, _____, _____

(b) 1, 3, 9, 27, _____, _____

5 (a) Circle the prime numbers in the list below.

21 37 47 51 63 71 81

(b) List the first five prime numbers.

____ ____ ____
____ ____

(c) Write down three prime factors of 42

____ ____ ____

6 (a) What temperature is 7 degrees lower than 4°C?

_____°C

(b) What number is 11 more than ⁻8 (negative 8)?

_____

7 Write down

(a) the square of 2 _____

(b) 2³ (the cube of 2) _____

(c) all of the square numbers between 30 and 70

_____

8 (a) Write **four thousand and eleven** in figures.

_____

(b) Write 14 908 in words. _____
_____

(c) Write in figures the Roman number MDCLXVIII.

_____

9 (a) What is the value of the 7 in 17 030?

_____

(b) Draw beads on the abacus to represent the number which is 10 more than 30 490

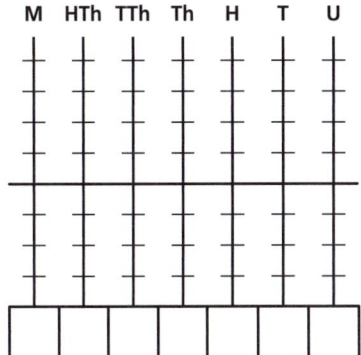

10 (a) Multiply 4090 by 10 _____

(b) Divide 70 578 by 100 _____

Number test 4

Final score

20 ___ %

1 (a) If these numbers are arranged in order of size, which will be in the middle?
   _____

   729 927 279 972 792

   (b) Write down the number which is exactly half way between 43 and 53
   _____

2 (a) On the scale, mark the positions of 7 and ⁻3

   (b) On the scale, mark the positions of 0.5 and 1.2

3 (a) On the scale, mark the position of 7.5

   (b) On the scale, mark the position of 80

   0 ─────────────── 100

4 (a) Round 4055 to the nearest 10 _____

   (b) Round 394 549 to the nearest 100
   _____

5 (a) Round 7.61 to the nearest whole number.
   _____

   (b) Round 7.072 to 1 decimal place.
   _____

6 (a) Shade a quarter of this shape.

   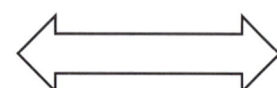

   (b) Complete these equivalent fractions:

   $\frac{1}{3} = \frac{}{9}$

   (c) Write these fractions in order of increasing size: $\frac{2}{3}, \frac{11}{12}, \frac{3}{4}$

   _____ _____ _____

   (d) Write $1\frac{1}{3}$ as an improper fraction. _____

7 (a) Add: $\frac{2}{7} + \frac{1}{7}$

   _____

   (b) Subtract: $\frac{4}{5} - \frac{2}{5}$

   _____

8 (a) Multiply: $\frac{1}{3} \times 12$

   _____

   (b) Divide: $\frac{2}{7} \div 2$

   _____

9 (a) Write the fraction $\frac{29}{100}$ as a decimal.
   _____

   (b) Write the decimal 0.17 as a fraction.
   _____

   (c) Write 20% as a fraction.
   _____

   (d) Write 95% as a decimal.
   _____

10 What is
   (a) 10% of 70 kg _____ kg

   (b) $\frac{3}{5}$ of £10? £_____

Number test 5

Final score
20 ___ %

1 (a) On the number grid, shade all the multiples of 3 and circle all the multiples of 5

31	32	33	34	35	36
41	42	43	44	45	46
51	52	53	54	55	56
61	62	63	64	65	66

(b) Write down a number between 10 and 20 which is a multiple of both 3 and 5
_____

2 (a) What is the largest number which is a factor of both 24 and 36? _____

(b) Complete the factor rainbow to show all of the factor pairs for 45

1 _____ _____ _____ _____ 45

3 (a) Count forwards in 100s from 1700

1700 _____ _____
_____ _____

(b) Count backwards in 10s from 1020

1020 _____ _____
_____ _____

4 Write the next two terms in each sequence.

(a) 1, 7, 13, 19, _____, _____

(b) 100, 91, 82, 73, _____, _____

5 (a) Circle the prime numbers in the list below.

15 27 33 43 57 67 77

(b) List the prime numbers between 30 and 40
_____

(c) Write down three prime factors of 30
_____

6 (a) What temperature is 5 degrees higher than ⁻7 °C?

_____ °C

(b) What number is 11 lower than 8?
_____

7 Write down

(a) the square of 7 _____

(b) 1³ (the cube of 1) _____

(c) all of the square numbers between 40 and 90
_____

8 (a) Write **thirty thousand, two hundred and seventeen** in figures.
_____

(b) Write 13 400 in words. _____
_____

(c) Write in figures the Roman number CCCLXXXVII.
_____

9 (a) What is the value of the 6 in 25 600?
_____

(b) Draw beads on the abacus to represent the number which is 1 more than 989 989

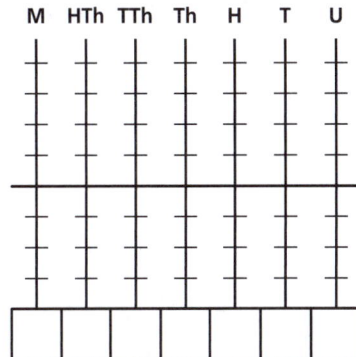

10 (a) Multiply 7905 by 10 _____

(b) Divide 45 700 by 1000 _____

11

Number test 6

Final score

20 ___ %

1 (a) If these numbers are arranged in order of size, which will be in the middle?

_____

2351   2315   2153   2135   2513

(b) Write down the number which is exactly half way between 28 and 54

_____

2 (a) On the scale, mark the positions of ⁻4 and 8

(b) On the scale, mark the positions of 0.4 and 1.1

3 (a) On the scale, mark the position of 3

0 —————————————— 10

(b) On the scale, mark the position of 90

0 —————————————— 100

4 (a) Round 3069 to the nearest 100

_____

(b) Round 74 890 to the nearest 1000

_____

5 (a) Round 1.499 to the nearest whole number. _____

(b) Round 12.47 to 1 decimal place.

_____

6 (a) Shade $\frac{1}{6}$ of this shape.

(b) Complete these equivalent fractions:

$\frac{2}{5} = \frac{}{15}$

(c) Write these fractions in order of increasing size: $\frac{2}{5}, \frac{3}{10}, \frac{7}{20}$

_____ _____ _____

(d) Write $\frac{5}{2}$ as a mixed number.

_____

7 (a) Add: $\frac{2}{9} + \frac{5}{9}$

_____

(b) Subtract: $\frac{5}{8} - \frac{1}{8}$

_____

8 (a) Multiply: $\frac{1}{5} \times 100$

_____

(b) Divide: $\frac{4}{5} \div 4$

_____

9 (a) Write the fraction $\frac{8}{100}$ as a decimal.

_____

(b) Write the decimal 0.28 as a fraction.

_____

(c) Write 80% as a fraction.

_____

(d) Write 10% as a decimal.

_____

10 What is
(a) 20% of 80 litres _____ litres
(b) $\frac{1}{4}$ of £18? £ _____

12

Number test 7

Final score
20 ___ %

1 (a) On the number grid, shade all the multiples of 6 and circle all the multiples of 7

44	45	46	47	48	49
54	55	56	57	58	59
64	65	66	67	68	69
74	75	76	77	78	79

(b) Write down a number between 80 and 90 which is a multiple of both 6 and 7

_____

2 (a) What is the largest number which is a factor of both 28 and 48?

_____

(b) Complete the factor rainbow to show all of the factor pairs for 54

1 ___ ___ ___ ___ ___ ___ 54

3 (a) Count forwards in 1000s from 8900

8900 _____ _____ _____ _____

(b) Count backwards in 100s from 3150

3150 _____ _____ _____ _____

4 Write the next two terms in each sequence.

(a) 1, 6, 11, 16, _____, _____

(b) 1100, 1020, 940, 860, _____, _____

5 (a) Circle the prime numbers in the list below.

23   39   43   59   63   79   81

(b) List the prime numbers between 0 and 20

_____

(c) Write down two prime factors of 45

_____   _____

6 (a) What temperature is 12 degrees lower than 11 °C?

_____ °C

(b) What number is 13 higher than ⁻7?

_____

7 Write down

(a) the square of 9 _____

(b) 4^3 (the cube of 4) _____

(c) all of the square numbers between 20 and 80

_____

8 (a) Write **two hundred thousand, two hundred and two** in figures.

_____

(b) Write 45 454 in words. _____

_____

(c) Write in figures the Roman number DCXI.

_____

9 (a) What is the value of the 9 in 190 000?

_____

(b) Draw beads on the abacus to represent the number which is 130 more than 4770

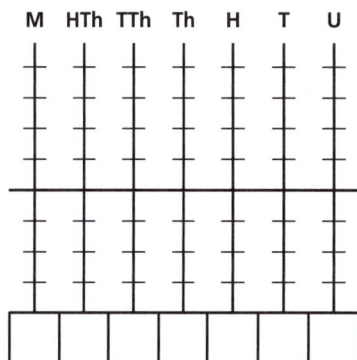

10 (a) Multiply 550 by 100 _____

(b) Divide 7700 by 1000 _____

Number test 8

Final score
20 ___ %

1 (a) If these numbers are arranged in order of size, which will be in the middle? _____

   1203   2130   312   1032   1230

(b) Write down the number which is exactly half way between 16 and 42 _____

2 (a) On the scale, mark the positions of ⁻3 and 5.5

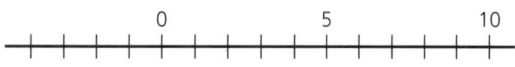

(b) On the scale, mark the positions of 0.8 and 1.25

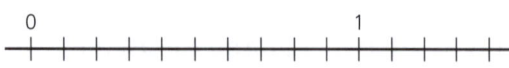

3 (a) On the scale, mark the position of 2

(b) On the scale, mark the position of 60

4 (a) Round 4935 to the nearest 10 _____

(b) Round 849 490 to the nearest 10 000 _____

5 (a) Round 5.55 to the nearest whole number. _____

(b) Round 17.61 to 1 decimal place. _____

6 (a) Shade $\frac{1}{2}$ of this shape.

(b) Complete these equivalent fractions:

$\frac{3}{7} = \frac{9}{}$

(c) Write these fractions in order of increasing size: $\frac{3}{10}, \frac{1}{8}, \frac{2}{9}$

_____ _____ _____

(d) Write $3\frac{1}{4}$ as an improper fraction. _____

7 (a) Add: $\frac{3}{8} + \frac{5}{8}$ _____

(b) Subtract: $\frac{4}{5} - \frac{1}{5}$ _____

8 (a) Multiply: $\frac{3}{4} \times 200$ _____

(b) Divide: $\frac{6}{7} \div 2$ _____

9 (a) Write the fraction $\frac{7}{100}$ as a decimal. _____

(b) Write the decimal 0.13 as a fraction. _____

(c) Write 40% as a fraction. _____

(d) Write 95% as a decimal. _____

10 What is

(a) 20% of 500 tonnes _____ tonnes

(b) $\frac{2}{3}$ of £24? £ _____

Number test 9

Final score
20 ___ %

1 (a) On the number grid, shade all the multiples of 6 and circle all the multiples of 11

64	65	66	67	68	69
74	75	76	77	78	79
84	85	86	87	88	89
94	95	96	97	98	99

(b) Write down a number between 100 and 140 which is a multiple of both 6 and 11
_____

2 (a) What is the largest number which is a factor of both 32 and 48? _____

(b) Complete the factor rainbow to show all of the factor pairs for 72

1 2 __ __ __ __ __ __ __ __ __ 72

3 (a) Count forwards in 100s from 19 900
19 900
_____ _____

_____ _____

(b) Count backwards in 10s from 1030
1030
_____ _____ _____ _____

4 Write the next two terms in each sequence.

(a) 1, 10, 19, 28, _____, _____

(b) 1, 2, 4, 7, 11, _____, _____

5 (a) Circle the prime numbers in the list below.

23 37 43 57 63 77 83

(b) Which number between 90 and 100 is prime?
_____

(c) Write down two prime factors of 50
_____ _____

6 (a) What temperature is 4 degrees lower than ⁻8°C?

_____ °C

(b) What number is 13 more than ⁻11 (negative 11)?

_____

7 Write down

(a) the square of 12 _____

(b) 5^3 (the cube of 5) _____

(c) all of the square numbers below 110

8 (a) Write **seventeen thousand and seventeen** in figures.

_____

(b) Write 40 404 in words. _____
_____

(c) Write in figures the Roman number MMDCLXV.

_____

9 (a) What is the value of the 1 in 210 450?

_____

(b) Draw beads on the abacus to represent the number which is 111 more than 189 989

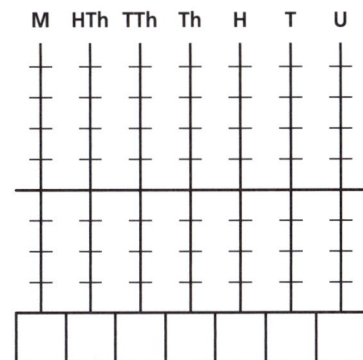

10 (a) Multiply 3303 by 100 _____

(b) Divide 70 000 by 100 000 _____

15

Number test 10

Final score

20 ___ %

1. (a) If these numbers are arranged in order of size, which will be in the middle?

    _____

    608   80.6   68   86   8.6

   (b) Write down the number which is exactly half way between 74 and 100

    _____

2. (a) On the scale, mark the positions of 6.5 and ⁻0.5

   (b) On the scale, mark the positions of ⁻0.15 and 1.3

3. (a) On the scale, mark the position of 4

   (b) On the scale, mark the position of 5

4. (a) Round 34 490 to the nearest 1000

    _____

   (b) Round 3 590 000 to the nearest million.

    _____

5. (a) Round 38.9 to the nearest whole number.

    _____

   (b) Round 4.04 to 1 decimal place.

    _____

6. (a) Shade a third of this shape.

   (b) Complete these equivalent fractions:

    $\frac{2}{5} = \frac{10}{}$

   (c) Write these fractions in order of increasing size: $\frac{2}{3}, \frac{4}{5}, \frac{3}{4}$

    _____   _____   _____

   (d) Write $\frac{13}{5}$ as an mixed number. _____

7. (a) Add: $\frac{5}{13} + \frac{9}{13}$

    _____

   (b) Subtract: $\frac{8}{9} - \frac{5}{9}$

    _____

8. (a) Multiply: $\frac{2}{5} \times 70$

    _____

   (b) Divide: $\frac{1}{6} \div 3$

    _____

9. (a) Write the fraction $\frac{23}{50}$ as a decimal.

    _____

   (b) Write the decimal 1.03 as a fraction.

    _____

   (c) Write 45% as a fraction.

    _____

   (d) Write 0.5% as a decimal.

    _____

10. What is:

    (a) 15% of 40 litres _____ litres

    (b) $\frac{3}{8}$ of £480? £_____

16

Calculations test 1

Final score

20 ___ %

Calculators must not be used in this test.

1 Write down the

  (a) sum of 12 and 5 _____

  (b) difference between 12 and 5 _____

  (c) product of 12 and 5 _____

  (d) remainder when 12 is divided by 5 _____

2 Complete the addition square.

+	3	8	4	7
6				
2				
9				
5				

3 Complete the multiplication square.

×	3	8	4	7
6				
2				
9				
5				

4 Write the other three addition and subtraction facts using the same numbers.

  31 − 7 = 24   _____ − _____ = _____

  _____ + _____ = _____

  _____ + _____ = _____

5 Write the other three multiplication and division facts using the same numbers.

  48 ÷ 6 = 8   _____ ÷ _____ = _____

  _____ × _____ = _____

  _____ × _____ = _____

6 Calculate:

  (a) 2 + 4 × 3

  _____

  (b) 8 × 7 − (6 + 4)

  _____

Calculate in your head, without written working.

7 (a) What is the sum of the first five odd numbers?

  _____

  (b) Subtract 99 from 500 _____

8 (a) How many ice creams priced at 99p could you buy with a £5 note?

  _____

  (b) Find the total cost of 5 chocolate bars priced at £1.99 each.

  £ _____

9 Given that 13 × 17 = 221, complete the following:

  (a) 13 × 170 = _____

  (b) 26 × 17 = _____

  (c) 4420 ÷ 13 = _____

Set out your working clearly.

10 (a) Add: 794 + 906

  _____

  (b) Add: 6.8 + 7.9

  _____

17

Calculations test 2

Final score

20 ___ %

1 (a) Subtract: 866 − 658

_____

(b) Subtract: 10.9 − 4.5

_____

2 (a) Multiply: 34 × 5

_____

(b) Multiply: 713 × 4

_____

3 (a) Multiply: 195 × 11

_____

(b) Multiply: 78 × 15

_____

4 (a) Divide: 1035 ÷ 3

_____

(b) Divide: 1428 ÷ 7

_____

5 (a) Divide, leaving a remainder: 200 ÷ 3

_____

(b) Divide, giving your answer as a decimal: 478 ÷ 4

_____

6 (a) Divide, giving your answer as a mixed number: 12 ÷ 5

_____

(b) Divide, using factors: 768 ÷ 16

_____

7 (a) Sam multiplied 46 by 5 to find the cost of 5 chocolate bars costing 46 pence each. The result was 230

What was the cost in pounds?

£ _____

(b) Sarah divided 76 by 8 to find how much 8 people would each receive when they shared a £76 prize. Her result was 9.5

How much did each person receive?

£ _____

8 The result of a calculation was 7.5

What would this mean in

(a) hours and minutes

_____ hours _____ minutes

(b) feet and inches?

_____ feet _____ inches

9 Complete these statements using **even** or **odd**:

(a) An even number times an odd number gives an _____ number.

(b) An odd number minus an odd number gives an _____ number.

10 Jane multiplied 497 by 11

(a) What is the units digit of the correct result?

_____

(b) By rounding both numbers, write down an approximate value for the result.

_____

18

Calculations test 3

Final score

20 ____ %

Calculators must not be used in this test.

1 Write down the

(a) sum of 17 and 26 _____

(b) difference between 17 and 26 _____

(c) product of 17 and 6 _____

(d) remainder when 23 is divided by 4 _____

2 Complete the subtraction square.

−	5	9	6	8
8	3	−1	2	
10	5			
7				
9				1

3 Complete the multiplication square.

×	4	9	5	8
7				
3				
8				
6				

4 Write the other three addition and subtraction facts using the same numbers.

44 − 19 = 25   ____ − ____ = ____

____ + ____ = ____

____ + ____ = ____

5 Write the other three multiplication and division facts using the same numbers.

72 ÷ 8 = 9   ____ ÷ ____ = ____

____ × ____ = ____

____ × ____ = ____

6 Calculate:

(a) 11 + 5 × 8

_____

(b) 9 × 6 − (9 − 6)

_____

Calculate in your head, without written working.

7 (a) What is the sum of the first five even numbers?

_____

(b) Subtract 408 from 1000 _____

8 (a) How many lollies priced at £1.08 could you buy with a £10 note?

_____

(b) Find the total cost of 5 packets of crisps priced at 49p each.

£ _____

9 Given that 29 × 11 = 319, complete the following:

(a) 22 × 29 = _____

(b) 290 × 11 = _____

(c) 31 900 ÷ 110 = _____

Set out your working clearly.

10 (a) Add: 348 + 899

_____

(b) Add: 4.3 + 19.8

_____

19

Calculations test 4

Final score
20 ___ %

1 (a) Subtract: 654 − 456

___

(b) Subtract: 18.5 − 7.8

___

2 (a) Multiply: 47 × 6

___

(b) Multiply: 294 × 5

___

3 (a) Multiply: 317 × 12

___

(b) Multiply: 56 × 17

___

4 (a) Divide: 2109 ÷ 3

___

(b) Divide: 4200 ÷ 6

___

5 (a) Divide, leaving a remainder: 120 ÷ 7

___

(b) Divide, giving your answer as a decimal: 376 ÷ 5

___

6 (a) Divide, giving your answer as a mixed number: 17 ÷ 4

___

(b) Divide, using factors: 800 ÷ 32

___

7 (a) Carrie multiplied $\frac{1}{4}$ by 15 to find the number of pizzas she should buy in order to give each of her friends a quarter of a pizza. The result was $3\frac{3}{4}$.

How many pizzas should she buy?

___

(b) Leo divided 48 by 5 to find how many sweets each boy would get if 5 boys shared a bag of 48 sweets.

Leo's result was 9.6, so he gave each boy 9 sweets. How many sweets were left?

___

8 The result of a calculation was 5.1

What would this mean in

(a) hours and minutes

___ hours ___ minutes

(b) metres and centimetres?

___ m ___ cm

9 Complete these statements using **even** or **odd**:

(a) An even number times an even number gives an ___ number.

(b) An even number minus an ___ number gives an even number.

10 Samantha multiplied 995 by 31

(a) What is the units digit of the correct result? ___

(b) By rounding both numbers, write down an approximate value for the result.

___

20

Calculations test 5

Final score

20 ___ %

Calculators must not be used in this test.

1 Write down the

   (a) sum of 33 and 11 _____

   (b) difference between 35 and 17 _____

   (c) product of 11 and 10 _____

   (d) remainder when 20 is divided by 7 _____

2 Complete the addition square.

+	9	5	8	6
7				
4				
12				
8				

3 Complete the multiplication square.

×	9	5	8	6
7				
4				
12				
8				

4 Write the other three addition and subtraction facts using the same numbers.

   28 − 19 = 9   _____ − _____ = _____

   _____ + _____ = _____

   _____ + _____ = _____

5 Write the other three multiplication and division facts using the same numbers.

   60 ÷ 12 = 5   _____ ÷ _____ = _____

   _____ × _____ = _____

   _____ × _____ = _____

6 Calculate:

   (a) (8 + 3) × 3

   _____

   (b) 8 × 7 − (6 − 4)

   _____

Calculate in your head, without written working.

7 (a) What is the sum of the first four square numbers?

   _____

   (b) Subtract 514 from 2000 _____

8 (a) How many muffins priced at 51p could you buy with a £10 note?

   _____

   (b) Find the total cost of 6 books priced at £4.99 each.

   £_____

9 Given that 45 × 29 = 1305, complete the following:

   (a) 90 × 29 = _____

   (b) 450 × 29 = _____

   (c) 13 050 ÷ 29 = _____

Set out your working clearly.

10 (a) Add: 1095 + 707

   _____

   (b) Add: 13.5 + 7.6

   _____

21

Calculations test 6

Final score
20 ___ %

1 (a) Subtract: 2345 – 518

_____

(b) Subtract: 16.4 – 9.7

_____

2 (a) Multiply: 57 × 6

_____

(b) Multiply: 108 × 9

_____

3 (a) Multiply: 238 × 12

_____

(b) Multiply: 54 × 16

_____

4 (a) Divide: 2004 ÷ 6

_____

(b) Divide: 4085 ÷ 5

_____

5 (a) Divide, leaving a remainder: 300 ÷ 7

_____

(b) Divide, giving your answer as a decimal: 500 ÷ 8

_____

6 (a) Divide, giving your answer as a mixed number: 20 ÷ 9

_____

(b) Divide, using factors: 448 ÷ 14

_____

7 (a) Crackers are sold in boxes of 6 and Morgan calculated that they would need exactly 15.5 boxes of crackers for the school Christmas party.

How many boxes should they buy and how many crackers will be left over?

_____ boxes and _____ crackers left over

(b) Kevin divided 2000 by 12 to find how many stamps each person would receive when 12 people shared a collection of 2000 stamps. His result was 166 remainder 8

What does the 8 represent? _____

_____

8 The result of a calculation was $2\frac{1}{4}$.

What would this mean in

(a) hours and minutes

_____ hours _____ minutes

(b) feet and inches?

_____ feet _____ inches

9 Complete these statements using **even** or **odd**:

(a) An even number minus an odd number gives an _____ number.

(b) An odd number multiplied by an _____ number gives an odd number.

10 Colin multiplied 1002 by 39

(a) What is the units digit of the correct result?

_____

(b) By rounding both numbers, write down an approximate value for the result.

_____

22

Calculations test 7

Final score

20 ___ %

Calculators must not be used in this test.

1   Write down the

  (a) sum of 43 and 37 _____

  (b) difference between 45 and 29 _____

  (c) product of 23 and 4 _____

  (d) remainder when 31 is divided by 7 _____

2   Complete the subtraction square.

−	3	8	9	5
9	6	1		
12	9			
8				
6		−2		

3   Complete the multiplication square.

×	7	6	8	9
8				
6				
11				
7				

4   Write the other three addition and subtraction facts using the same numbers.

56 − 39 = 17   _____ − _____ = _____

_____ + _____ = _____

_____ + _____ = _____

5   Write the other three multiplication and division facts using the same numbers.

132 ÷ 11 = 12   _____ ÷ _____ = _____

_____ × _____ = _____

_____ × _____ = _____

6   Calculate:

  (a) 9 × 6 + 31

  _____

  (b) 7 × 6 − 4 × 6

  _____

Calculate in your head, without written working.

7 (a) What is the sum of the numbers from 10 to 13 inclusive?

  _____

  (b) Subtract 348 from 500 _____

8 (a) How many packets of stamps priced at 50p could you buy for £13.50?

  _____

  (b) Find the total cost of 6 bags of sweets priced at 75p each.

  £ _____

9   Given that 47 × 83 = 3901, complete the following:

  (a) 47 × 830 = _____

  (b) 94 × 83 = _____

  (c) 390 100 ÷ 83 = _____

Set out your working clearly.

10 (a) Add: 4503 + 497

  _____

  (b) Add: 18.4 + 26.5

  _____

23

Calculations test 8

Final score
20 ___ %

1 (a) Subtract: 723 – 456

   _____

   (b) Subtract: 20.5 – 11.8

   _____

2 (a) Multiply: 89 × 7

   _____

   (b) Multiply: 607 × 8

   _____

3 (a) Multiply: 715 × 11

   _____

   (b) Multiply: 303 × 13

   _____

4 (a) Divide: 5040 ÷ 9

   _____

   (b) Divide: 1001 ÷ 7

   _____

5 (a) Divide, leaving a remainder: 200 ÷ 6

   _____

   (b) Divide, giving your answer as a decimal: 150 ÷ 4

   _____

6 (a) Divide, giving your answer as a mixed number: 29 ÷ 5

   _____

   (b) Divide, using factors: 126 ÷ 18

   _____

7 (a) Wood shelving comes only in 2.4 metre lengths. Ali calculated that he needed about 18 metres of shelving. He divided 18 by 2.4 and the result was 7.5

   How many 2.4 m lengths should he buy?

   _____

   (b) Clare and 7 friends won £100 and Clare divided 100 by 8 to see how much each would get. Her result was 12 remainder 4

   How much would each person receive?

   £_____

8 The result of a calculation was 1.6

   What would this mean in

   (a) hours and minutes

   _____ hours _____ minutes

   (b) kilograms and grams?

   _____ kg _____ g

9 Complete these statements using **even** or **odd**:

   (a) An odd number times an _____ number gives an odd number.

   (b) An odd number plus an _____ number gives an odd number.

10 Philippa multiplied 603 by 89

   (a) What is the units digit of the correct result? _____

   (b) By rounding both numbers, write down an approximate value for the result.

   _____

24

Calculations test 9

Final score

20 ___ %

Calculators must not be used in this test.

1 Write down the

(a) sum of 73 and 27 _____

(b) difference between 37 and 45 _____

(c) product of 11 and 12 _____

(d) remainder when 28 is divided by 5 _____

2 Complete the addition square.

+	9	11	6	8
7				
8				
12				
6				

3 Complete the multiplication square.

×	9	11	6	8
7				
8				
12				
6				

4 Write the other three addition and subtraction facts using the same numbers.

67 − 49 = 18   _____ − _____ = _____

_____ + _____ = _____

_____ + _____ = _____

5 Write the other three multiplication and division facts using the same numbers.

72 ÷ 8 = 9   _____ ÷ _____ = _____

_____ × _____ = _____

_____ × _____ = _____

6 Calculate:

(a) $(7 + 4) \times (2 + 5)$

_____

(b) $9 \times 8 − (6 + 7)$

_____

Calculate in your head, without written working.

7 (a) What is the product of the first five integers?

_____

(b) Subtract £4.89 from £10.00 £_____

8 (a) How many biscuits priced at 21p could you buy if you had £2.45 in coins?

_____

(b) Find the total cost of 7 DVDs priced at £7.99 each.

£_____

9 Given that $48 \times 53 = 2544$, complete the following:

(a) $53 \times 480 =$ _____

(b) $24 \times 106 =$ _____

(c) $2544 \div 212 =$ _____

Set out your working clearly.

10 (a) Add: 3965 + 509

_____

(b) Add: 17.4 + 86.7

_____

25

Calculations test 10

Final score
20 ___ %

1 **(a)** Subtract: 1012 – 821

___

(b) Subtract: 15.6 – 7.7

___

2 **(a)** Multiply: 68 × 7

___

(b) Multiply: 312 × 8

___

3 **(a)** Multiply: 408 × 12

___

(b) Multiply: 82 × 19

___

4 **(a)** Divide: 1040 ÷ 8

___

(b) Divide: 1107 ÷ 9

___

5 **(a)** Divide, leaving a remainder: 150 ÷ 7

___

(b) Divide, giving your answer as a decimal: 396 ÷ 8

___

6 **(a)** Divide, giving your answer as a mixed number: 19 ÷ 8

___

(b) Divide, using factors: 7254 ÷ 18

___

7 **(a)** When Toni calculated the number of pounds of potatoes that she needed to make chips for her friends, she got 4.25

What is this in pounds and ounces?

_____ pounds _____ ounces

(b) Sandy divided 144 by 20 to find how many marbles each 20 friends would get when they shared a gross of marbles. His result was 7.2

How many marbles were left over?

___

8 The result of a calculation was 2.05

What would this mean in

(a) hours and minutes

_____ hours _____ minutes

(b) centimetres and millimetres?

_____ cm _____ mm

9 Complete these statements using **even** or **odd**.

(a) If an even number divides exactly by an odd number the result is an _____ number.

(b) An odd number minus an even number gives an _____ number.

10 Jane multiplied 196 by 9.5

(a) What is the **tenths** digit of the correct result? ___

(b) By rounding both numbers, write down an approximate value for the result.

___

26

Problem solving test 1

Final score
20 ___ %

Questions 1 to 3 concern this pattern sequence.

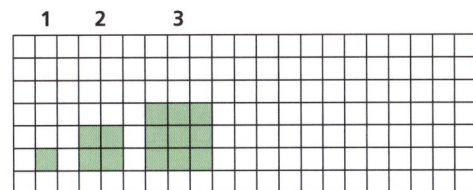

1 Draw pattern **4** and pattern **5** in the sequence.

2 Complete this table of data.

Pattern number	1	2	3	4	5
Area (square units)		4			
Perimeter (units)	4		12		

3 For pattern 6, what would be the

   (a) area _____ square units

   (b) perimeter? _____ units

Questions 4 and 5 concern Roman numerals.

4 (a) Which number, in Roman numerals, is shown on the abacus below?

   _____

   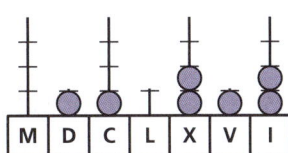

   (b) Write this number in words. _____
   _____

   (c) Write the number in figures. _____

5 (a) On the Roman abacus, why do the D, L and V spikes have space for only one bead?
   _____

   (b) Draw beads on the blank Roman abacus picture to show the number which is three more than the number on the abacus in question 4

   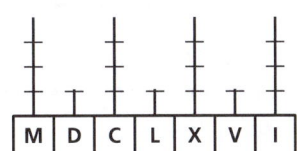

Questions 6 to 10 concern the five number cards below.

The cards can be placed side by side to make numbers. For example: the largest 5-digit number that can be made is

6 (a) What is the largest 4-digit even number that can be made? _____

   (b) What is the smallest 5-digit odd number that can be made? _____

7 (a) Write down the four of the 2-digit prime numbers that can be made using these cards.
   _____

   (b) Which 2-digit square number can be made?

   _____

The cards could be used in a different way using the operation signs: (+ − × ÷). For example:

5 × 8 + 3 = 43

8 Use the same three cards to make:

   (a) 37 _____

   (b) 29 _____

9 Use any three cards to make:

   (a) 32 _____

   (b) 28 _____

10 (a) Which three cards have a sum of 16?

   _____ _____ _____

   (b) Which three cards have a product of 70?

   _____ _____ _____

27

Problem solving test 2

Final score 20 ___ %

1. John buys 2 chocolate bars priced at 64p each and 2 bags of crisps priced at 46p each.

   (a) How much change will John receive from a £5 note? £_____

   (b) John receives 4 different coins in change. List the 4 coins. _____

2. The nutrition information on a chocolate biscuit is:

	per 20 g biscuit
Fat	5 g
Carbohydrate	12 g
Protein	1 g

   (a) What fraction of a biscuit is fat? _____

   (b) What percentage of a biscuit is protein? _____%

3. The pupils in Year 5 are going to a concert. They travel in 4 minibuses, which can each carry 17 people including the driver. Each minibus is driven by a teacher and there is 1 extra teacher on each minibus. There are 5 spare seats altogether.

   How many pupils are there in Year 5? _____

4. Helen is twice as old as her brother Jamie. The sum of their ages is 18 years.

   (a) How old is Helen? _____

   (b) How many years older than Jamie will Helen be 10 years from now?

   _____ years

5. The diagram below represents a 12 cm by 4 cm rectangle.

   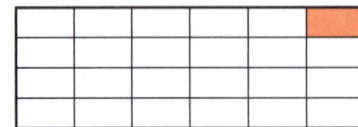

   (a) What is the perimeter of the red rectangle? _____ cm

   (b) What is the area of the red rectangle? _____ cm²

6. Shade, as accurately as possible

    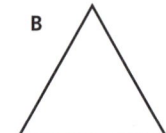

   (a) a third of the equilateral triangle **A**

   (b) a quarter of the equilateral triangle **B**.

7. The table below records the hours of sunshine in the five days of Jo's holiday.

Day	Mon	Tue	Wed	Thu	Fri
Hours of sun	8	7	5	6	9

   (a) How many hours of sunshine did Jo enjoy altogether on her holiday? _____ hours

   (b) What was the average number of hours of sunshine on a day? _____ hours

8. The Carroll diagram below shows the favourite sports of the pupils in Year 5

	Hockey	Football
Boys	9	13
Girls	16	8

   (a) How many pupils are in Year 5? _____

   (b) How many more pupils prefer hockey than prefer football? _____

   (c) What fraction of the girls prefer hockey? _____

9. Jasmine has thought of two numbers, both less than 20. The product of the numbers is 65

   What is the sum of the numbers? _____

10. The 10:55 bus goes direct from Ayburn to Cetown.

Ayburn	09:23	10:55
Beford	10:14	....
Cetown	10:30	11:49

    How many minutes quicker is the journey from Ayburn to Cetown by the 10:55 bus? _____ minutes

28

Problem solving test 3

Final score

20 ___ %

Questions 1 to 3 concern this pattern sequence.

1   2   3

1 Draw pattern **4** and pattern **5** in the sequence.

2 Complete this table of data.

Pattern number	1	2	3	4	5
Number of triangles	1	3			
Perimeter (units)			7		

3 For pattern 6, what would be the

(a) number of triangles _____

(b) perimeter? _____ units

Questions 4 and 5 concern applying a rule, over and over again, to 2-digit numbers.

The rule is: Multiply the tens digit by 3 and then add the units digit. Stop when you reach a single-digit number. For example:

25 → 11 → 4
61 → 19 → 12 → 5
99 → 36 → 15 → 8

4 Show what happens when the rule is applied to:

(a) 13 _____

(b) 29 _____

(c) 37 _____

5 Find a number between 40 and 50 which reaches the single-digit number 7 _____

Questions 6 to 10 concern the five number cards below.

The cards can be placed side by side to make numbers. For example: the smallest 4-digit number that can be made is

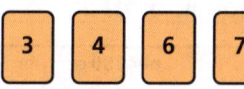

6 (a) What is the largest 4-digit even number that can be made?

_____

(b) What is the largest 5-digit odd number that can be made?

_____

7 (a) Write down three 2-digit square numbers that can be made using these cards.

_____ _____ _____

(b) Which 2-digit cube number can be made?

_____

The cards could be used in a different way using the operation signs: (+ − × ÷). For example:

3 × 4 + 6 = 18

8 Use **the same three** cards to make:

(a) 22 _____

(b) 21 _____

9 Use **any two** cards to make:

(a) 11 _____

(b) 28 _____

10 (a) Which **four** cards have a sum of 20?

_____ _____ _____ _____

(b) Which **three** cards have a product of 84?

_____ _____ _____

Problem solving test 4

Final score
$\overline{20}$ ___ %

1 Sara pays for 2 packets of sweets priced at 90p each and 2 bags of crisps with a £5 note. She receives £2.30 in change.

   (a) What is the price of a bag of crisps?
   _____ p

   (b) The £2.30 change is given in 3 coins.
   List the 3 coins. _____

2 The nutrition information on a trifle is:

	per 100 g	per 150 g serving
Fat	8 g	_____ g
Carbohydrate	18 g	_____ g
Protein	2 g	_____ g
Fibre	1 g	_____ g

   (a) Complete the table for a 150 g serving.

   (b) What percentage of a 150 g helping of trifle is protein? _____%

3 The diagram shows information about the favourite pets of children in Year 5

	Dogs	Cats
Girls	13	15
Boys	23	11

   (a) How many children in Year 5 said that dogs were their favourite pet? _____

   (b) How many more boys are there than girls in Year 5? _____

4 (a) George is 2 years older than his brother Jon. The sum of their ages is 20 years.
   How old is George? _____

   (b) Jon's birthday is 26 July and George's birthday is 10 days later.
   What date is George's birthday?
   _____

5 A frame in the shape of a 12 cm cube is made from pieces of stiff wire.
   What is the total length of wire used? _____ cm

6 The shape below is made by joining two identical equilateral triangles.

   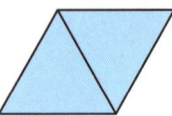

   (a) Name the shape. _____

   (b) Name a shape which can be made by joining three equilateral triangles.
   _____

7 The table below records the money earned by five friends for doing cleaning jobs.

Name	Amy	Ben	Cris	Dan	Eva
Amount earned (£)	4.50	3.60	5.50	6.40	5.00

   (a) How much was earned altogether? £_____

   The friends decided to share the money equally between them.

   (b) How much money did each friend receive? £_____

8 Here are the times of films on TV channels A and B.

	A	B
Start	19:45	19:50
End	21:35	21:45

   Which film is the longer and by how many minutes? Channel _____ by _____ minutes

9 In a word-making game, a letter with reflection symmetry scores 2 points and all other letters score 1 point.

   What would be the score for each of the following words?

   (a) SYMMETRY _____ points

   (b) ABACUS _____ points

   (c) BEHAVE _____ points

10 Iain has thought of two whole numbers. The sum of the numbers is 25 and the difference between them is 7

   (a) What are the numbers? _____ and _____

   (b) What is the product of the numbers? _____

Problem solving test 5

Final score: 20 ___ %

Questions 1 to 3 concern this pattern sequence.

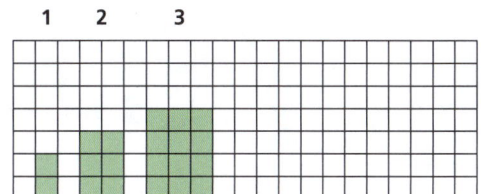

1 Draw pattern **4** and pattern **5** in the sequence.

2 Complete this table of data.

Pattern number	1	2	3	4	5
Area (square units)		6			
Perimeter (units)	6		14		

3 For pattern 6, what would be the

   (a) area _____ square units

   (b) perimeter? _____ units

Questions 4 and 5 concern the first eleven prime numbers:

2 3 5 7 11 13 17 19 23 29 31

4 (a) Write down a pair of prime numbers which have a sum of:

   10 = _____ + _____

   20 = _____ + _____

   (b) Write down a pair of prime numbers which have a sum of:

   30 = _____ + _____

   40 = _____ + _____

5 Not counting 2 and 3, look at the sum of pairs of consecutive (next to each other) primes.

   5 + 7 = **12**   7 + 11 = **18**   11 + 13 = **24**

   (a) Complete the following: 13 + _____ = **30**
       17 + 19 = _____   19 + _____ = **42**

   (b) Complete the statement below.

   The **bold** numbers are all multiples of _____

   It would be exciting if this pattern continued, but 23 + 29 = 52 and we'd like it to be **48**!

   (c) Writing prime numbers in the spaces, complete the following:

   _____ + _____ = **48**

   _____ + _____ = **54**

   _____ + _____ = **60**

Questions 6 to 10 concern the 8 cards below.

Number cards:    Operation cards: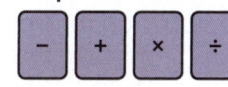

The cards can be placed side by side to make number sentences. For example:

[5] [+] [3] = 8   [2] [4] [×] [5] = 120

6 Write the result to make each of these number sentences correct:

   (a) [3] [4] [−] [2] [5] = _____

   (b) [5] [×] [3] [+] [4] = _____

   (c) [4] [×] [5] [−] [2] [3] = _____

7 Show how the cards could be put together to make the results below. *Just write the numbers and signs. Don't draw the cards!*

   (a) 13 _____

   (b) 14 _____

8 Show how the cards can be put together to make 59 _____

9 Using brackets, show how all 4 number cards can be put together to make 48

   _____

10 Sarah says that the largest number you can make is 5432 without using any operation cards. Freya says that the largest number is the result of 52 × 43

   (a) Write an approximate result for Freya's idea.

   _____

   (b) Who is correct?

   _____

31

Problem solving test 6

Final score
20 ___ %

1 Julian buys 4 muffins priced at 95p each and 4 mugs of coffee priced at £1.15 each.

(a) How much change will Julian receive from a £10 note? £_____

(b) Julian receives 5 coins in change.

List one possible combination.
_____

2 The nutrition information on a tart is:

	per 100 g
Fat	15 g
Carbohydrate	45 g
Protein	5 g

One slice of the tart has a mass of 80 g.

What is the mass of fat in a slice of tart?
_____ g

3 Complete the Venn diagram below using the information given about the members of Year 5

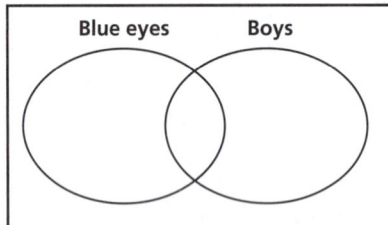

- 12 girls do not have blue eyes.
- Half of the girls have blue eyes.
- There are 60 members in Year 5
- $\frac{1}{4}$ of the boys have blue eyes.

4 A TV programme starts at 19:55 and lasts for 1 hour and 40 minutes.

(a) At what time does it end? _____ : _____

During the programme there are seven 3-minute breaks for adverts.

(b) What percentage of the programme time is taken up by adverts? _____ %

5 A rectangular patio is made up of square slabs. It is 9 slabs long and 4 slabs wide. If the slabs are lifted and rearranged as a square, how many slabs will there be along each edge of the square? _____

6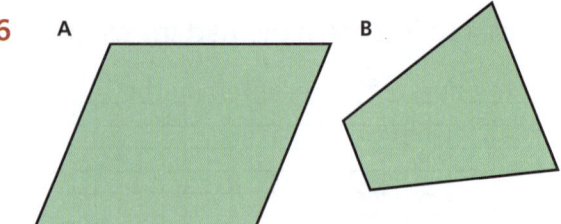

A   B

(a) Draw lines on shape A to cut it into 4 identical pieces.

(b) Draw one line on shape B to cut it into two pieces of equal area.

7 The table below records the numbers of sweets eaten one week by Alf and Bel.

	Su	Mo	Tu	We	Th	Fr	Sa
Alf	3	2	5	4	5	3	3
Bel	4	5	3	5	1	5	4

Who ate most sweets during the week?
_____

8 The block graph below records the colours of cars in a showroom.

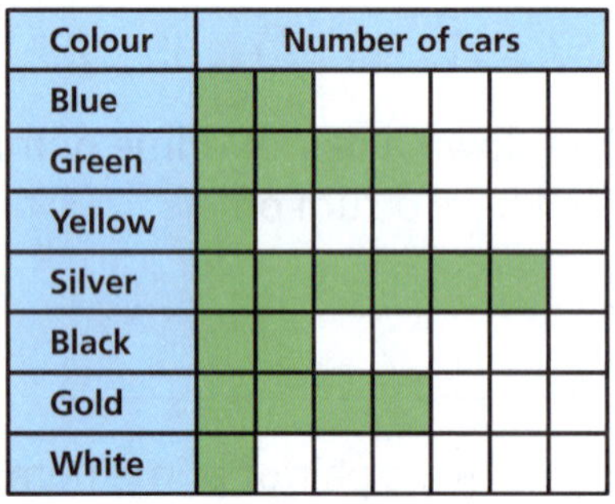

Colour	Number of cars
Blue	■■
Green	■■■
Yellow	■
Silver	■■■■
Black	■■
Gold	■■■
White	■

One block represents one car.

What percentage of the cars is silver?
_____ %

9 Lisa has thought of two numbers, both less than 20

The product of the numbers is 77

What is the sum of the numbers? _____

10 Write the numbers 1, 2, 3, 4, 5, 6, 7 and 8 in this grid so that each column has the same sum and each row has the same sum.

Problem solving test 7

Final score
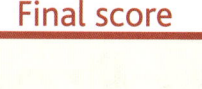
20 ___ %

Questions 1 to 3 concern this pattern sequence.

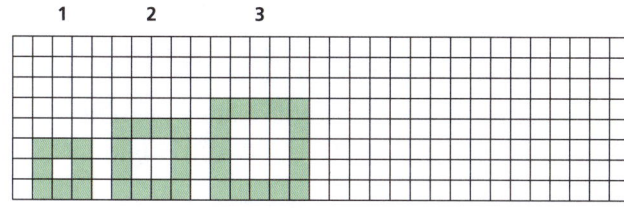

1   Draw pattern **4** in the sequence.

2   Complete this table of data.

Pattern number	1	2	3	4
Number of white squares	1	4		
Number of green squares	8		16	
Total number of squares			25	

3   (a) How many green squares are in pattern 5?

   _____

   (b) What is the total number of squares in pattern 6?

   _____

Questions 4 and 5 concern applying a rule, over and over again, to 2-digit starting numbers.

The rule is: Double the units digit and then add the tens digit. Stop when you reach a single-digit number. For example:

25 → 12 → 5

68 → 22 → 6

99 → 27 → 16 → 13 → 7

4   Show what happens when the rule is applied to:

   (a) 47 _____

   (b) 56 _____

   (c) 19 _____

5   Find a number between 10 and 20 which reaches the single-digit number 9

   _____

Questions 6 to 10 concern the five number cards below.

The cards can be placed side by side to make numbers. For example: the largest 3-digit odd number that can be made is

6   (a) What is the largest 4-digit even number that can be made?

   _____

   (b) What is the largest 5-digit odd number that can be made?

   _____

7   (a) Write down the largest 2-digit multiple of 3 that can be made using these cards.

   _____

   (b) Which 2-digit square number can be made?

   _____

The cards could be used in a different way using the operation signs: (+ − × ÷). For example:

5 × 6 + 7 = 37

8   Use **the same three** cards to make

   (a) 18 _____

   (b) 4 _____

9   Use **any three** cards to make

   (a) 25 _____

   (b) 39 _____

10 (a) Which **four** cards have a sum of 24?

   _____

   (b) Which **three** cards have a product of 240?

   _____

Problem solving test 8

Final score
20 ___ %

1. Harry pays for 3 concert tickets priced at £4.50 each and 3 bottles of juice with a £20 note. He receives £4.70 in change.

   What is the price of a bottle of juice?
   _____ p

2. The nutrition information on a 'ready meal' is:

	per 100 g	per 500 g meal
Fat	4 g	_____ g
Carbohydrate	12 g	_____ g
Protein	9 g	_____ g

   (a) Complete the masses for a 500 g meal.

   The recommended adult daily intake of protein is 50 g.

   (b) What percentage of the daily protein intake is provided by this meal?
   _____%

3. Complete the diagram below from the information given about the members of a swimming club.

   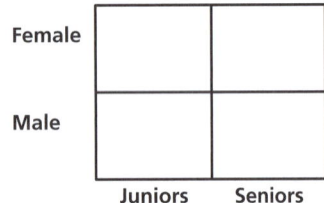

   - The club has 40 members.
   - $\frac{3}{4}$ of the members are juniors.
   - There are 10 more females than males.
   - There are 17 junior girls.

4. Julius Caesar was born in July 100 BC.

   How old was he when he was killed in March 44 BC? _____

5. The diagram shows a path of width 1 m, round a square lawn which has sides of length 7 m.

   What is the area of the path? _____ m²

6. Jasmine has two identical isosceles triangles.

   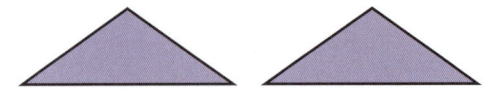

   Name the three different shapes Jasmine could make by joining the two triangles:

   A with no lines of symmetry _____

   B with 1 line of symmetry _____

   C with 2 lines of symmetry. _____

7. The table below records the heights (in cm) and masses (in kg) of six friends.

Name	Hal	Ian	Jon	Ken	Lee	Mel
Height	135	129	141	138	132	140
Mass	41	35	40	37	39	36

   (a) If the friends were arranged in order of increasing height, who would be standing next to Ken? _____ and _____

   (b) How many of the friends are heavier than Mel? _____

8. The timetable shows times for two buses, X and Y. Both buses take the same time between the towns. Complete the timetable.

	X	Y
Ayby	08:45	09:30
Beton	09:30	____:____
Ceford	11:22	____:____

9. Sukrit is thinking of a number and has given these clues. The number is:
   - less than 50
   - 1 more than a square number
   - prime.

   Suggest four possibilities for Sukrit's number.
   _____ or _____ or _____ or _____

10. Fido eats 2 sachets of dog food each day. A box of 12 sachets costs £5.00

    What is the cost of feeding Fido for 360 days? £_____

34

Problem solving test 9

Final score
20 ___ %

Questions 1 to 3 concern this pattern sequence.

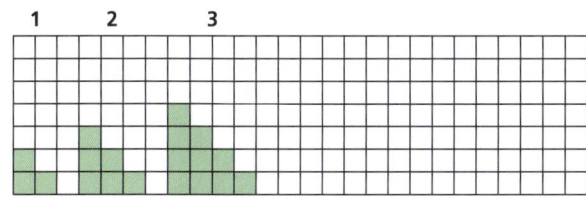

1 Draw pattern **4** in the sequence.

2 Complete this table of data.

Pattern number	1	2	3	4	5
Area (square units)		6			
Perimeter (units)	8		16		

3 For pattern 6, what would be the

(a) area _____ square units

(b) perimeter? _____ units

Questions 4 and 5 concern the sum of the digits of numbers. For example:

12, 102, 111 and 1002 all have a digit sum of 3

4 (a) Write down all of the numbers less than 100 that have a digit sum of 3

   _____

  (b) Write down all of the numbers between 100 and 300 that have a digit sum of 3

   _____

5 (a) Write down, in order, all of the numbers less than 100 that have a digit sum of 4

   _____

  (b) Write down all of the numbers between 100 and 300 that have a digit sum of 5

   _____

  (c) What is the next largest number after 600 that has a digit sum of 6?

   _____

Questions 6 to 10 concern the 8 cards below.

Number cards   Operation cards

The cards can be placed side by side to make number sentences. For example:

6 × 7 = 42   6 7 + 8 9 = 156

6 Write the result to make each of these number sentences correct:

(a) 9 6 − 8 7 = _____

(b) 6 × 7 + 8 = _____

(c) 6 + 7 × 8 = _____

7 Show how the cards could be put together to make the results below. *Just write the numbers and signs. Don't draw the cards!*

(a) 4 _____

(b) 5 _____

8 Show how the cards can be put together to make

(a) 11 _____

(b) ⁻2 _____

9 Show how just two cards can be used to make a fraction equivalent to $\frac{3}{4}$. _____

10 Kelly says that 96 × 87 gives the largest result. Liam says that 97 × 86 gives the largest result and correctly calculates 8342

(a) Calculate 96 × 87 _____

(b) Who is correct? _____

35

Problem solving test 10

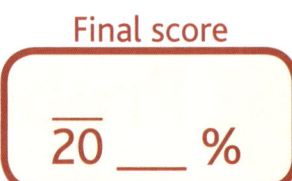

Final score
20 ___ %

1. Chris buys 7 scones priced at 80p each, 4 cups of tea priced at 65p each and 3 cups of coffee priced at 95p each.

   (a) What is the total cost? £_____

   Chris pays the exact amount in the smallest possible number of notes and coins.

   (b) Which notes and coins does he hand over? _____

2. The nutrition information on a cereal bar is:

	per 100 g	per bar
Fat	15 g	5 g
Carbohydrate	60 g	20 g
Protein	6 g	2 g

   (a) What percentage of a bar is fat? _____%

   (b) What is the approximate mass of one bar? _____ g

3. The Venn diagram below shows information about the musical activities of pupils in a school.

   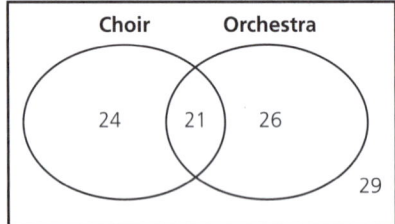

   (a) How many pupils are in the school? _____

   (b) How many pupils are **not** in the choir? _____

   (c) What percentage of the pupils are in the orchestra? _____%

4. Hannah came second in a cross-country running race with a time of 37 minutes 48 seconds. The winner beat her by 1 minute 13 seconds. What was the winner's time?
   _____ minutes _____ seconds

5. Jane has cut from card a rectangle measuring 12 cm by 7 cm and a square with sides of 9 cm. Which shape has the larger area and by how many square centimetres?
   _____ by _____ cm²

6. (a) Draw a line to show how the shape **A** can be cut to form an isosceles triangle and an isosceles trapezium.

   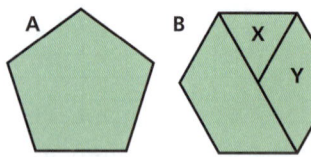

   (b) Shape **B** has been cut into 3 pieces.

   Name the two smaller pieces, **X** and **Y**.
   X _____
   Y _____

7. The table records the temperatures, in °C, in three towns at midday, one week in January.

Town	Su	Mo	Tu	We	Th	Fr	Sa
Aytown	−3	−2	−1	−3	−3	0	1
Betown	−2	0	−1	0	1	−1	2
Cetown	0	−1	−3	1	1	2	1

   (a) On Wednesday, how many degrees warmer was Cetown than Aytown?
   _____ degrees

   (b) Which town had the largest range of midday temperatures? _____

8. The pictogram records ice-cream sales.

   One symbol 🍦 represents **two** ice-creams.

Vanilla	🍦🍦🍦🍦🍦
Chocolate	🍦🍦🍦🍦
Mint	🍦🍦🍦🍦🍦🍦🍦
Strawberry	🍦🍦🍦

   What was the total number of ice-creams sold? _____

9. Nicole has thought of a 2-digit number. The tens digit is 1 more than the units digit and the number is a multiple of 7

   What is Nicole's number? _____

10. How many 50 cm square carpet tiles would be needed to cover a floor measuring 5 m by 4 m? _____

Pre-algebra test 1

Final score
20 ___ %

Questions 1 to 5 concern missing numbers.

1 Find the number represented by the symbol in each of these sentences.

  (a) $5 + \bullet = 13$ $\bullet =$ _____

  (b) $\ast - 7 = 16$ $\ast =$ _____

  (c) $\diamond + \diamond = 32$ $\diamond =$ _____

2 Find the number represented by the symbol in each of these sentences.

  (a) $5 \times \bullet = 30$ $\bullet =$ _____

  (b) $\ast \times 7 = 35$ $\ast =$ _____

  (c) $\diamond \times \diamond = 64$ $\diamond =$ _____

3 Find the number represented by the symbol in each of these sentences.

  (a) $15 + \bullet = 113$ $\bullet =$ _____

  (b) $\ast - 17 = 6$ $\ast =$ _____

  (c) $50 - \diamond = 32$ $\diamond =$ _____

4 Find the number represented by the symbol in each of these sentences.

  (a) $15 \times \bullet = 45$ $\bullet =$ _____

  (b) $\ast + 8 = 6$ $\ast =$ _____

  (c) $20 - \diamond = \diamond$ $\diamond =$ _____

5 Find the number represented by the symbol in each of these sentences.

  (a) $9 + \bullet = 17 - \bullet$ $\bullet =$ _____

  (b) $\ast \div 4 = 8$ $\ast =$ _____

  (c) $\frac{1}{2} \diamond = 18$ $\diamond =$ _____

Questions 6 to 8 concern missing operation signs.

6 On each line write the operation sign ($- + \times \div$) to make the sentences true.

  (a) 4 _____ $7 = 11$

  (b) 24 _____ $4 = 4$ _____ 2

  (c) 8 _____ 2 _____ $5 = 11$

7 In each of these, write the operation signs ($- + \times \div$) on the lines, to make the sentences true.

  (a) 4 _____ $8 = 2$ _____ 10

  (b) 14 _____ $3 = 7$ _____ 6

  (c) $(11$ _____ $3)$ _____ $7 = 40$

8 In each of these, write the operation signs ($- + \times \div$) on the lines, to make the sentences true.

  (a) 2 _____ 7 _____ $3 = 42$

  (b) 4 _____ $4 = 12$ _____ 4

  (c) $(5$ _____ $3)$ _____ $(2$ _____ $4) = 48$

Questions 9 and 10 concern simple word formulae.

9 The perimeter of an equilateral triangle can be found using the word formula:

Multiply the length of one side by three.

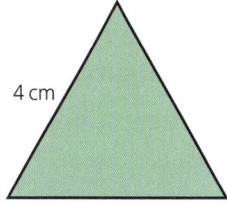

4 cm

  (a) What is the perimeter of the equilateral triangle above? _____ cm

  (b) What would be the length of a side of an equilateral triangle with perimeter 24 cm? _____ cm

10 The volume of a cuboid can be found using the word formula:

Multiply the length by the width by the height.

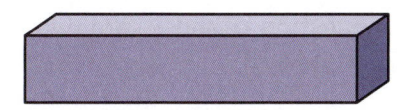

  (a) What is the volume of a cuboid which is 10 cm long, 5 cm wide and 4 cm high? _____ cm³

  (b) What would be the height of a cuboid which is 6 cm long, 4 cm wide and has volume 96 cm³? _____ cm

Pre-algebra test 2

Final score
20 ___ %

1. Lengths in centimetres can be converted into lengths in millimetres by using the simple word formula:

   Multiply by ten.

   Write a similar word formula which can be used to convert lengths in millimetres into lengths in centimetres.

   _____

Questions 2 to 5 concern number puzzles.

2. (a) William was asked to think of a number and he chose 11

   He was asked to multiply his number by 4 and write the result.

   What did he write? _____

   (b) William was asked to think of another number and perform the operation × 4

   This time he wrote down 120

   What number did he think of? _____

3. Freya was given the number 6 and asked to perform an operation on it. 24 was the result.

   Suggest two different operations that Freya might have done.

   _____ or _____

4. Iain has some marbles in his pocket but he does not know how many there are.

   He takes 10 marbles from his pocket to play a game with Kiera and Liam. He wins 4 marbles from Kiera and loses 11 to Liam.

   He finds that he now has a total of 23 marbles. How many marbles did Iain have in his pocket before the game? _____

5. In this question ❖ and ☞ are two integers (whole numbers).

   ❖ − ☞ = 7 and ❖ + ☞ = 17

   What are ❖ and ☞ ? ❖ = _____ and ☞ = _____

Questions 6 and 7 concern sequences.

6. (a) Draw the next 4 beads in this sequence.

   ●●○●●○●●

   (b) Draw the next 4 beads in this sequence.

   ●○●○●○●○●○●○●○●○●○

7. Study this sequence of shape beads.

   □○◇○□□○◇○□□○◇

   The 1st bead is a square and the 3rd bead is a diamond.

   What will be the shape of

   (a) the 15th bead _____

   (b) the 50th bead _____

   (c) the 51st bead? _____

Questions 8 to 10 concern number machines.

8. Tammy has a machine which adds 3 to any number that she puts into it.

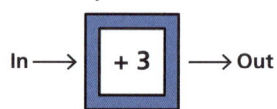

   In ⟶ [+3] ⟶ Out

   (a) If Tammy puts in 7, what number comes out? _____

   (b) If 23 comes out, what number did Tammy put in? _____

9. David has a machine, but the label has fallen off and he can't remember what it does!

   In ⟶ [ ] ⟶ Out

   If David puts in 2, 4 comes out. If he puts in 4, 6 comes out.

   Write the operation on the label.

10. On the grid, the crosses represent the (in, out) numbers (2, 4) and (4, 6) for David's machine.

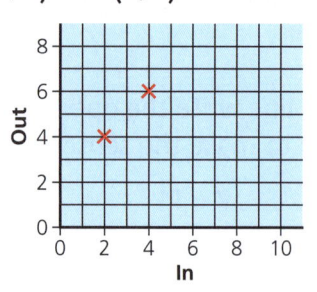

    Mark with crosses four more (in, out) pairs of numbers for David's machine.

38

Answers

Number test 1

1 (a)

4	5	⑥	7	8	⑨
14	⑮	16	17	⑱	19
㉔	25	26	㉗	28	29
34	35	㊱	37	38	㊴

(b) 42 48
2 (a) 6 (b) 3 4 6
3 (a) 1100 1200 1300 1400
  (b) 210 200 190 180
4 (a) 17, 21 (b) 16, 32
5 (a) 23 and 61 circled
  (b) 11 13 17 19
  (c) 2 3
6 (a) −1 °C (b) 4
7 (a) 16 (b) 27 (c) 16 25
8 (a) 524
  (b) three thousand and fifty-six
  (c) 776
9 (a) 8000
  (b) abacus showing 508 080 (M HTh TTh Th H T U: _ 5 0 8 0 8 0)
10 (a) 3450 (b) 70.8

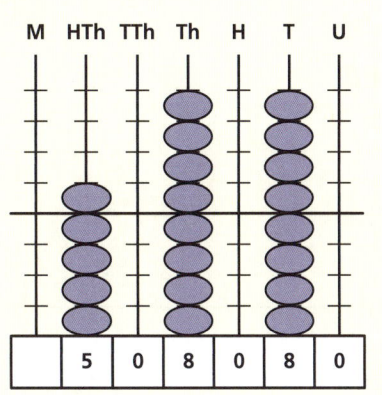

Number test 2

1 (a) 453 (b) 26
2 (a) number line with −2, 0, 4, 5, 10
  (b) number line with 0, 0.2, 1, 1.3
3 (a) number line 0, 5, 10
  (b) number line 0, 25, 100
4 (a) 7500 (b) 910 000
5 (a) 5 (b) 17.6
6 (a) any quarter of shape shaded (b) $\frac{2}{8}$
  (c) $\frac{1}{12}$ $\frac{1}{8}$ $\frac{1}{4}$ (d) $\frac{3}{2}$
7 (a) $\frac{4}{5}$ (b) $\frac{2}{7}$
8 (a) 2 (b) $\frac{1}{8}$
9 (a) 0.35 (b) $\frac{43}{100}$
  (c) $\frac{49}{100}$ (d) 0.65
10 (a) 65 g (b) £80

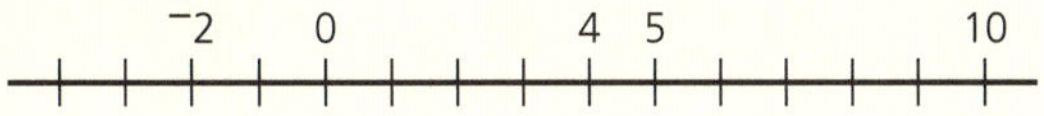

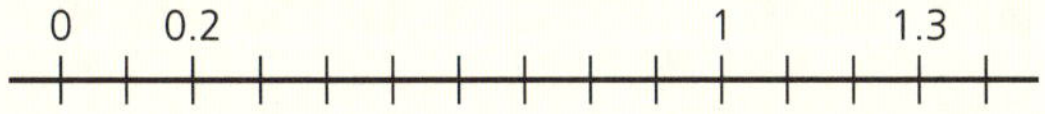

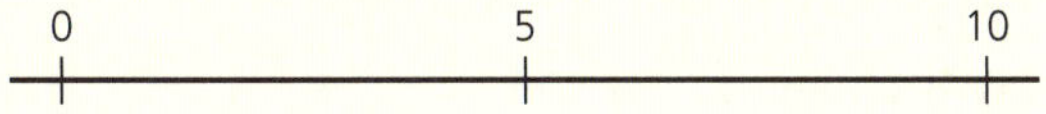

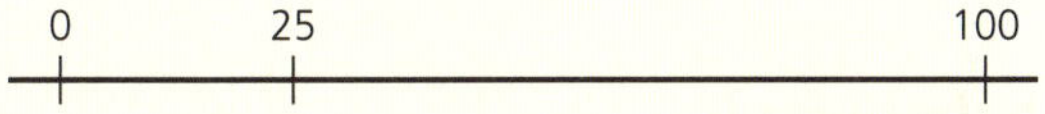

Number test 3

1 (a)

1	2	3	④	5	6
11	⑫	13	14	15	⑯
21	22	23	㉔	25	26
31	㉜	33	34	35	㊱

(b) 48
2 (a) 15 (b) 4 7 14
3 (a) 985 995 1005 1015
  (b) 940 840 740 640
4 (a) 13, 16 (b) 81, 243
5 (a) 37 47 71 circled
  (b) 2 3 5 7 11
  (c) Any two of: 2 3 7
6 (a) −3 °C (b) 3
7 (a) 4 (b) 8
  (c) 36 49 64
8 (a) 4011
  (b) fourteen thousand, nine hundred and eight
  (c) 1668
9 (a) 7000
  (b) abacus showing 30 500 (M HTh TTh Th H T U: _ _ _ 3 0 5 0 0)
10 (a) 40 900 (b) 705.78

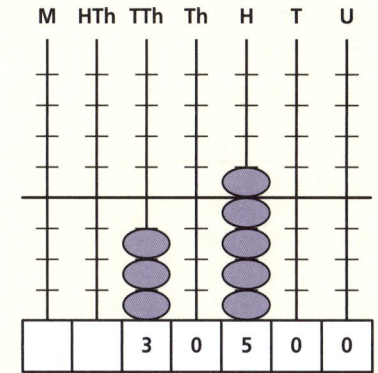

Number test 4

1 (a) 792 (b) 48
2 (a) number line with −3, 0, 5, 7, 10
  (b) number line with 0, 0.5, 1, 1.2
3 (a) number line 0, 7.5, 10
  (b) number line 0, 80, 100
4 (a) 4060 (b) 394 500
5 (a) 8 (b) 7.1
6 (a) any quarter of shape shaded
  (b) $\frac{3}{9}$ (c) $\frac{2}{3}$ $\frac{3}{4}$ $\frac{11}{12}$
  (d) $\frac{4}{3}$
7 (a) $\frac{3}{7}$ (b) $\frac{2}{5}$
8 (a) 4 (b) $\frac{1}{7}$
9 (a) 0.29 (b) $\frac{17}{100}$
  (c) $\frac{1}{5}$ (d) 0.95
10 (a) 7 kg (b) £6

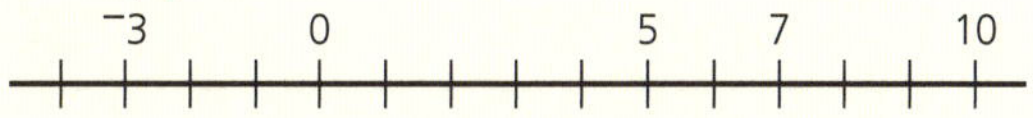

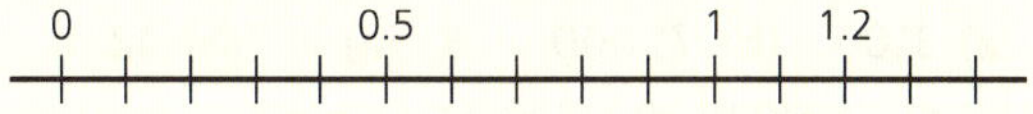

Number test 5

1 (a)
31	32	**33**	34	㉟	36
41	**42**	43	44	㊺	46
51	52	**53**	**54**	㊺	56
61	62	**63**	64	㊺	66

(b) 15

2 (a) 12 (b) 3 5 9 15
3 (a) 1800 1900 2000 2100
    (b) 1010 1000 990 980
4 (a) 25, 31 (b) 64, 55
5 (a) 43 67 circled (b) 31 37 (c) 2 3 5
6 (a) −2 °C (b) −3
7 (a) 49 (b) 1 (c) 49 64 81
8 (a) 30 217
    (b) thirteen thousand four hundred (c) 387
9 (a) 600
    (b) M HTh TTh Th H T U — 9 8 9 9 9 0
10 (a) 79 050 (b) 45.7

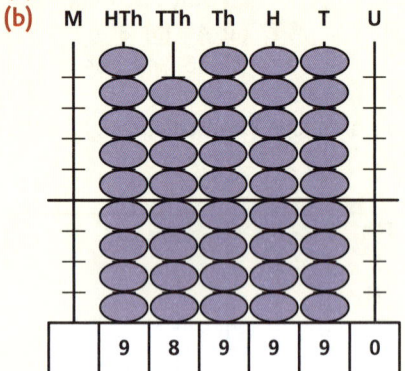

Number test 6

1 (a) 2315 (b) 41
2 (a) number line −4 to 10
    (b) number line 0 to 1.1
3 (a) number line 0 to 10 (3 marked)
    (b) number line 0 to 100 (90 marked)
4 (a) 3100 (b) 75 000
5 (a) 1 (b) 12.5
6 (a) any sixth of shape shaded
    (b) $\frac{6}{15}$ (c) $\frac{3}{10}$ $\frac{7}{20}$ $\frac{2}{5}$ (d) $2\frac{1}{2}$
7 (a) $\frac{7}{9}$ (b) $\frac{1}{2}$
8 (a) 20 (b) $\frac{1}{5}$
9 (a) 0.08 (b) $\frac{7}{25}$ (c) $\frac{4}{5}$ (d) 0.1
10 (a) 16 litres (b) £4.50

Number test 7

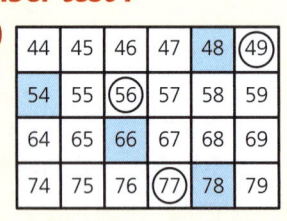

1 (a)
44	45	46	47	**48**	㊾
54	55	㊻	57	58	59
64	65	**66**	67	68	69
74	75	76	㊼	**78**	79

(b) 84

2 (a) 4 (b) 2 3 6 9 18 27
3 (a) 9900 10 900 11 900 12 900
    (b) 3050 2950 2850 2750
4 (a) 21, 26 (b) 780, 700
5 (a) 23 43 59 79 circled
    (b) 2 3 5 7 11 13 17 19 (c) 3 5
6 (a) −1 °C (b) 6
7 (a) 81 (b) 64 (c) 25 36 49 64
8 (a) 200 202
    (b) forty-five thousand, four hundred and fifty-four
    (c) 611
9 (a) 90 000
    (b) M HTh TTh Th H T U — 4 9 0 0
10 (a) 55 000 (b) 7.7

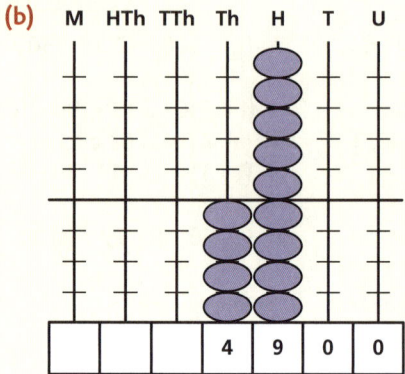

Number test 8

1 (a) 1203 (b) 29
2 (a) number line −3 to 10, 5.5 marked
    (b) number line 0 to 1.25
3 (a) number line 0 to 10 (2 marked)
    (b) number line 0 to 100 (60 marked)
4 (a) 4940 (b) 850 000
5 (a) 6 (b) 17.6
6 (a) any half of shape shaded
    (b) $\frac{9}{21}$ (c) $\frac{1}{8}$ $\frac{2}{9}$ $\frac{3}{10}$ (d) $\frac{13}{4}$
7 (a) 1 (b) $\frac{3}{5}$
8 (a) 150 (b) $\frac{3}{7}$
9 (a) 0.07 (b) $\frac{13}{100}$ (c) $\frac{2}{5}$ (d) 0.95
10 (a) 100 tonnes (b) £16

Number test 9

1 (a)
64	65	㊻	67	68	69
74	75	**76**	㊼	**78**	79
84	85	86	87	㊿	89
94	95	**96**	97	98	㊿

(b) 132

2 (a) 16 (b) 3 4 6 8 9 12 18 24 36

3 (a) 20 000 20 100 20 200 20 300
  (b) 1020 1010 1000 990
4 (a) 37, 46   (b) 16, 22
5 (a) 23 37 43 83 circled   (b) 97   (c) 2 5
6 (a) −12 °C   (b) 2
7 (a) 144   (b) 125
  (c) 1 4 9 16 25 36 49 64 81 100
8 (a) 17 017
  (b) forty thousand, four hundred and four   (c) 2665
9 (a) 10 000
  (b)

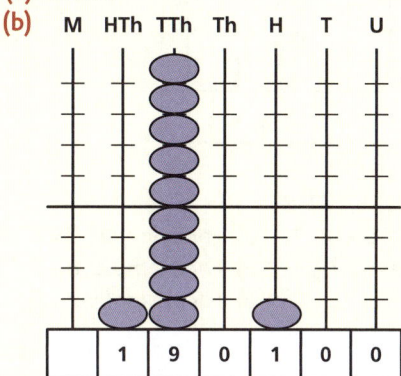

M	HTh	TTh	Th	H	T	U
	1	9	0	1	0	0

10 (a) 330 300   (b) 0.7

Number test 10

1 (a) 80.6   (b) 87
2 (a) number line showing −0.5, 0, 5, 6.5, 10
  (b) number line showing −0.15, 0, 1, 1.3
3 (a) 0 ... 4 ... 10
  (b) 0 5 ... 100
4 (a) 34 000   (b) 4 000 000   5 (a) 39   (b) 4.0
6 (a) any third of shape shaded
  (b) $\frac{10}{25}$ (c) $\frac{2}{3}$ $\frac{3}{4}$ $\frac{4}{5}$ (d) $2\frac{3}{5}$
7 (a) $1\frac{1}{13}$ (b) $\frac{1}{3}$ 8 (a) 28   (b) $\frac{1}{18}$
9 (a) 0.46   (b) $1\frac{3}{100}$ (c) $\frac{9}{20}$ (d) 0.005
10 (a) 6 litres   (b) £180

Calculations test 1

1 (a) 17   (b) 7   (c) 60   (d) 2
2

+	3	8	4	7
6	9	14	10	13
2	5	10	6	9
9	12	17	13	16
5	8	13	9	12

3

×	3	8	4	7
6	18	48	24	42
2	6	16	8	14
9	27	72	36	63
5	15	40	20	35

4 31 − 24 = 7   7 + 24 = 31   24 + 7 = 31
5 48 ÷ 8 = 6   6 × 8 = 48   8 × 6 = 48
6 (a) 14   (b) 46   7 (a) 25   (b) 401
8 (a) 5   (b) £9.95   9 (a) 2210   (b) 442
                       (c) 340
10 (a) 1700   (b) 14.7

Calculations test 2

1 (a) 208   (b) 6.4   2 (a) 170   (b) 2852
3 (a) 2145 (b) 1170   4 (a) 345   (b) 204
5 (a) 66 remainder 2   (b) 119.5
6 (a) $2\frac{2}{5}$ (b) 48   7 (a) £2.30   (b) £9.50
8 (a) 7 hours 30 minutes   (b) 7 feet 6 inches
9 (a) even   (b) even   10 (a) 7   (b) 5000

Calculations test 3

1 (a) 43   (b) 9   (c) 102   (d) 3
2

−	5	9	6	8
8	3	−1	2	0
10	5	1	4	2
7	2	−2	1	−1
9	4	0	3	1

3

×	4	9	5	8
7	28	63	35	56
3	12	27	15	24
8	32	72	40	64
6	24	54	30	48

4 44 − 25 = 19   19 + 25 = 44   25 + 19 = 44
5 72 ÷ 9 = 8   8 × 9 = 72   9 × 8 = 72
6 (a) 51   (b) 51   7 (a) 30   (b) 592
8 (a) 9   (b) £2.45   9 (a) 638   (b) 3190
                      (c) 290
10 (a) 1247   (b) 24.1

Calculations test 4

1 (a) 198   (b) 10.7   2 (a) 282   (b) 1470
3 (a) 3804 (b) 952   4 (a) 703   (b) 700
5 (a) 17 remainder 1   (b) 75.2
6 (a) $4\frac{1}{4}$ (b) 25   7 (a) 4   (b) 3
8 (a) 5 hours 6 minutes   (b) 5 m 10 cm
9 (a) even   (b) even   10 (a) 5   (b) 30 000

Calculations test 5

1 (a) 44   (b) 18   (c) 110   (d) 6
2

+	9	5	8	6
7	16	12	15	13
4	13	9	12	10
12	21	17	20	18
8	17	13	16	14

3

×	9	5	8	6
7	63	35	56	42
4	36	20	32	24
12	108	60	96	72
8	72	40	64	48

4 28 − 9 = 19   19 + 9 = 28   9 + 19 = 28
5 60 ÷ 5 = 12   5 × 12 = 60   12 × 5 = 60
6 (a) 33   (b) 54   7 (a) 30   (b) 1486
8 (a) 19   (b) £29.94   9 (a) 2610   (b) 13 050
10 (a) 1802   (b) 21.1   (c) 450

Calculations test 6

1 (a) 1827 (b) 6.7   2 (a) 342   (b) 972
3 (a) 2856 (b) 864   4 (a) 334   (b) 817
5 (a) 42 remainder 6   (b) 62.5
6 (a) $2\frac{2}{9}$ (b) 32
7 (a) 16 boxes and 3 crackers left   (b) 8 stamps left over
8 (a) 2 hours 15 minutes   (b) 2 feet 3 inches
9 (a) odd   (b) odd   10 (a) 8   (b) 40 000

Mathematics Workbook: 10-minute Maths Tests Age 8–10 published by Galore Park

Calculations test 7

1 (a) 80 (b) 16 (c) 92 (d) 3

2
−	3	8	9	5
9	6	1	0	4
12	9	4	3	7
8	5	0	−1	3
6	3	−2	−3	1

3
×	7	6	8	9
8	56	48	64	72
6	42	36	48	54
11	77	66	88	99
7	49	42	56	63

4 56 − 17 = 39   39 + 17 = 56   17 + 39 = 56
5 132 ÷ 12 = 11   11 × 12 = 132   12 × 11 = 132
6 (a) 85   (b) 18   7 (a) 46   (b) 152
8 (a) 27   (b) £4.50   9 (a) 39 010   (b) 7802
                               (c) 4700
10 (a) 5000   (b) 44.9

Calculations test 8

1 (a) 267 (b) 8.7   2 (a) 623 (b) 4856
3 (a) 7865 (b) 3939   4 (a) 560 (b) 143
5 (a) 33 remainder 2   (b) 37.5
6 (a) $5\frac{4}{5}$ (b) 7   7 (a) 8 (b) £12.50
8 (a) 1 hours 36 minutes   (b) 1 kg 600 g
9 (a) odd   (b) even   10 (a) 7   (b) 54 000

Calculations test 9

1 (a) 100 (b) 8 (c) 132 (d) 3

2
+	9	11	6	8
7	16	18	13	15
8	17	19	14	16
12	21	23	18	20
6	15	17	12	14

3
×	9	11	6	8
7	63	77	42	56
8	72	88	48	64
12	108	132	72	96
6	54	66	36	48

4 67 − 18 = 49   49 + 18 = 67   18 + 49 = 67
5 72 ÷ 9 = 8   8 × 9 = 72   9 × 8 = 72
6 (a) 77   (b) 59   7 (a) 120   (b) £5.11
8 (a) 11   (b) £55.93
9 (a) 25 440   (b) 2544   (c) 12
10 (a) 4474   (b) 104.1

Calculations test 10

1 (a) 191 (b) 7.9   2 (a) 476 (b) 2496
3 (a) 4896 (b) 1558   4 (a) 130 (b) 123
5 (a) 21 remainder 3   (b) 49.5
6 (a) $2\frac{3}{8}$ (b) 403   7 (a) 4 pounds 4 ounces
                                    (b) 4
8 (a) 2 hours 3 minutes   (b) 2 cm 0.5 mm
9 (a) even   (b) odd   10 (a) 0   (b) 2000

Problem solving test 1

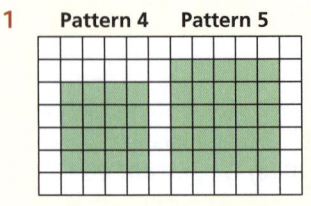

1 Pattern 4   Pattern 5

2
Pattern number	1	2	3	4	5
Area (square units)	1	4	9	16	25
Perimeter (units)	4	8	12	16	20

3 (a) 36 square units   (b) 24 units
4 (a) DCXXVII
  (b) six hundred and twenty-seven   (c) 627
5 (a) two Vs make an X and so on
  (b)

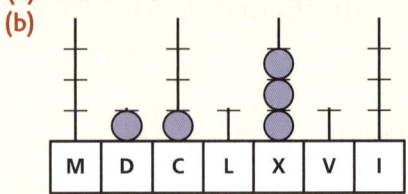

6 (a) 8752   (b) 23 587
7 (a) Any of : 23 37 53 73 83   (b) 25
8 (a) For example: 5 × 8 − 3   (b) For example: 3 × 8 + 5
9 (a) For example: 5 × 7 − 3   (b) For example: 8 ÷ 2 × 7
10 (a) For example: 3 5 8   (b) For example: 2 5 7

Problem solving test 2

1 (a) £2.80   (b) £2 50p 20p 10p
2 (a) $\frac{1}{4}$ (b) 5%   3 55
4 (a) 12   (b) 6 years   5 (a) 6 cm   (b) 2 cm²
6 (a) any third of shape shaded
  (b) any quarter of shape shaded
7 (a) 35 hours   (b) 7 hours
8 (a) 46   (b) 4   (c) $\frac{2}{3}$ 9 18
10 13 minutes

Problem solving test 3

1 Pattern 4   Pattern 5

2
Pattern number	1	2	3	4	5
Number of triangles	1	3	5	7	9
Perimeter (units)	3	5	7	9	11

3 (a) 11   (b) 13 units
4 (a) 13 → 6   (b) 29 → 15 → 8   (c) 37 → 16 → 9
5 42
6 (a) 9764   (b) 97 643   7 (a) 36 49 64 (b) 64
8 (a) For example: 3 × 6 + 4   (b) For example: 6 × 4 − 3
9 (a) For example: 7 + 4   (b) For example: 4 × 7
10 (a) For example: 3 4 6 7   (b) For example: 3 4 7

Problem solving test 4

1 (a) 45p   (b) £2 20p 10p
2 (a)

	per 100 g	per 150 g serving
Fat	8 g	12 g
Carbohydrate	18 g	27 g
Protein	2 g	3 g
Fibre	1 g	1.5 g

(b) 2%
3 (a) 36   (b) 6   4 (a) 11   (b) 5 August
5 144 cm

6 (a) rhombus (b) isosceles trapezium
7 (a) £25 (b) £5
8 Channel B by 5 minutes
9 (a) 14 points (b) 11 points (c) 12 points
10 (a) 9 and 16 (b) 144

Problem solving test 5

1 Pattern 4 Pattern 5

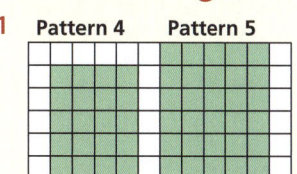

2
Pattern number	1	2	3	4	5
Area (square units)	2	6	12	20	30
Perimeter (units)	6	10	14	18	22

3 (a) 42 square units (b) 26 units
4 (a) $10 = 3 + 7$ $20 = 7 + 13$ or $17 + 3$
  (b) $30 = 13 + 17$ or $11 + 19$ or $23 + 7$
      $40 = 17 + 23$ or $11 + 29$
5 (a) $13 + 17 = 30$ $17 + 19 = 36$ $19 + 23 = 42$
  (b) 6 (c) $19 + 29 = 48$ $23 + 31 = 54$ $29 + 31 = 60$
6 (a) 9 (b) 19 (c) -3
7 (a) For example: $2 \times 5 + 3$ or $2 \times 4 + 5$ or $52 \div 4$
  (b) For example: $2 + 3 + 4 + 5$ or $4 \times 3 + 2$
8 For example: $24 + 35$ or $25 + 34$
9 For example: $(3 + 5) \times (4 + 2)$
10 (a) 2000 (b) Sarah

Problem solving test 6

1 (a) £1.60 (b) £1 20p 20p 10p 10p
             or 50p 50p 20p 20p 20p
             or 50p 50p 50p 5p 5p
2 12 g
3 Blue eyes / Boys Venn diagram: 12, 9, 27, 12

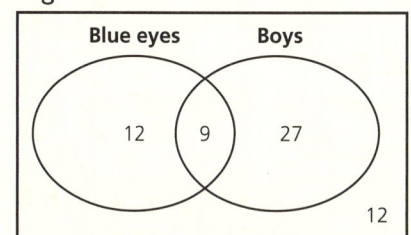

4 (a) 21:35 (b) 21%   5 6
6 (a) shape divided into four identical parallelograms
  (b) shape divided into two equal pieces
7 Bel   8 30%   9 18
10
1	7	6	4
8	2	3	5

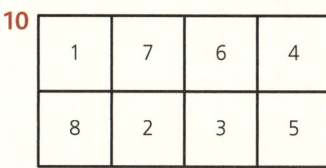

The row totals are both 18 and the column totals are all 9 – the sum of the first 8 numbers is 36; $4 \times 9 = 36$ and $2 \times 18 = 36$
Columns may be in different positions; whole rows may be reversed (top to bottom).

Problem solving test 7

1 Pattern 4

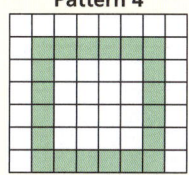

2
Pattern number	1	2	3	4
Number of white squares	1	4	9	16
Number of green squares	8	12	16	20
Total number of squares	9	16	25	36

3 (a) 24 (b) 64
4 (a) $47 \to 18 \to 17 \to 15 \to 11 \to 3$
  (b) $56 \to 17 \to 15 \to 11 \to 3$ (c) $19 \to 19 \to 19 \ldots$
5 14
6 (a) 8756 (b) 87 653   7 (a) 87 (b) 36
8 (a) For example: $5 + 6 + 7$ (b) For example: $5 + 6 - 7$
9 (a) For example: $6 \times 3 + 7$ or $(8 - 3) \times 5$
  (b) For example: $6 \times 7 - 3$ or $(8 + 5) \times 3$
10 (a) For example: $3 + 6 + 7 + 8$
   (b) For example: $5 \times 6 \times 8$

Problem solving test 8

1 (a) 60p
2 (a)
	per 100 g	per 500 g meal
Fat	4 g	20 g
Carbohydrate	12 g	60 g
Protein	9 g	45 g

  (b) 90%
3
	Juniors	Seniors
Female	17	8
Male	13	2

4 55
5 32 m²
6 A parallelogram B delta (arrowhead) kite C rhombus
7 (a) Hal and Mel (b) 4
8
	X	Y
Ayby	08:45	09:30
Beton	09:30	10:15
Ceford	11:22	12:07

9 2 or 5 or 17 or 37
10 £300

Problem solving test 9

1 Pattern 4

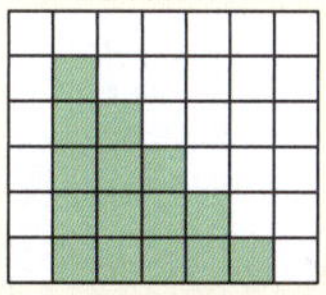

2
Pattern number	1	2	3	4	5
Area (square units)	3	6	10	15	21
Perimeter (units)	8	12	16	20	24

Mathematics Workbook: 10-minute Maths Tests Age 8–10 published by Galore Park

3 (a) 28 square units (b) 28 units
4 (a) 3 12 21 30 (b) 102 111 120 201 210 300
5 (a) 4 13 22 31 40
  (b) 104 113 122 131 140 203 212 221 230
  (c) 1005
6 (a) 9 (b) 50 (c) 62
7 (a) For example: 6 + 7 − 9
  (b) For example: 6 + 7 − 8 or 6 + 8 − 9
8 (a) For example: 79 − 68
  (b) For example: 7 − 9 or 6 − 8
9 For example: 6 ÷ 8   10 (a) 8352 (b) Kelly

Problem solving test 10

1 (a) £11.05 (b) £10 £1 5p
2 (a) 15% (b) 33 g
3 (a) 100 (b) 55 (c) 47%
4 36 minutes 35 seconds   5 rectangle by 3 cm²
6 (a) [pentagon figure]
  (b) X equilateral triangle Y rhombus
7 (a) 4 degrees
  (b) Cetown
8 40
9 21
10 80

Pre-algebra test 1

1 (a) ☞ = 8 (b) ✳ = 23 (c) ✦ = 16
2 (a) ☞ = 6 (b) ✳ = 5 (c) ✦ = 8
3 (a) ☞ = 98 (b) ✳ = 23 (c) ✦ = 18
4 (a) ☞ = 3 (b) ✳ = −2 (c) ✦ = 10
5 (a) ☞ = 4 (b) ✳ = 32 (c) ✦ = 36
6 (a) + (b) ÷ + (c) × −
7 (a) + + (b) × × (c) × +
8 (a) × × (b) × + or + − (c) + × +
9 (a) 12 cm (b) 8 cm 10 (a) 200 cm³ (b) 4 cm

Pre-algebra test 2

1 divide by ten   2 (a) 44 (b) 30
3 +18 or ×4   4 30   5 ✦ = 12 ☞ = 5
6 (a) ● ● ● ○
  (b) ● ○ ○ ○
7 (a) square (b) square (c) square
8 (a) 10 (b) 20   9 +2
10 four more points plotted, e.g. (1, 3), (3, 5), (6, 8)

Pre-algebra test 3

1 (a) ☞ = 16 (b) ✳ = 23 (c) ✦ = 13
2 (a) ☞ = 8 (b) ✳ = 9 (c) ✦ = 7
3 (a) ☞ = 15 (b) ✳ = 62 (c) ✦ = 47
4 (a) ☞ = 3 (b) ✳ = 23 (c) ✦ = 22
5 (a) ☞ = 12 (b) ✳ = 72 (c) ✦ = 72
6 (a) + (b) − × (c) + −
7 (a) + ÷ (b) × × (c) − ×
8 (a) × × (b) + − (c) + × ÷ or − × −
9 (a) 42 cm (b) 5 cm 10 (a) 36 cm³ (b) 3 cm

Pre-algebra test 4

1 multiply by a thousand
2 (a) 39 (b) 60 3 ÷3 or −14   4 27
5 ✦ = 18 ☞ = 6

6 (a) ● ● ○
  (b) red
7 (a) A (b) H (c) A   8 (a) 24 (b) 15
9 +4
10 (a) points (2, 6) and (3, 7) plotted
  (b) three more points plotted, eg. (1, 5), (4, 8), (5, 9),

Pre-algebra test 5

1 (a) ☞ = 18 (b) ✳ = 26 (c) ✦ = 21
2 (a) ☞ = 6 (b) ✳ = 9 (c) ✦ = 9
3 (a) ☞ = 18 (b) ✳ = 54 (c) ✦ = 17
4 (a) ☞ = 3 (b) ✳ = 34 (c) ✦ = 25
5 (a) ☞ = 12 (b) ✳ = 64 (c) ✦ = 16
6 (a) − (b) ÷ + (c) + +
7 (a) + − (b) ÷ + (c) + ×
8 (a) × × (b) ÷ ÷ (c) + × −
9 (a) 30 cm (b) 14 cm 10 (a) 60 cm³ (b) 5 cm

Pre-algebra test 6

1 divide by thirty 2 (a) 56 (b) 32 3 −16 or ÷3
4 2   5 ✦ = 20 ☞ = 10 6 11 14 17 20
7 (a) E (b) E 20 V 10
8 (a) 40 (b) 16   9 +3
10 (a) points (1, 4) and (3, 6) plotted
  (b) three more points plotted, e.g. (2, 5), (4, 7), (5, 8)

Pre-algebra test 7

1 (a) ☞ = 19 (b) ✳ = 36 (c) ✦ = 2½
2 (a) ☞ = 9 (b) ✳ = 8 (c) ✦ = 11
3 (a) ☞ = 211 (b) ✳ = 60 (c) ✦ = 23
4 (a) ☞ = 5 (b) ✳ = 27 (c) ✦ = 35
5 (a) ☞ = 3 (b) ✳ = 63 (c) ✦ = 8
6 (a) + (b) ÷ − (c) × ×
7 (a) + × (b) − × (c) + ×
8 (a) × ÷ (b) ÷ ÷ (c) + ÷ +
9 (a) 16 cm (b) 7 cm 10 (a) 330 cm³ (b) 8 cm

Pre-algebra test 8

1 divide by one hundred   2 (a) 65 (b) 108
3 ÷2 or −16   4 46   5 ✦ = 3 ☞ = 8
              or ✦ = 8 ☞ = 3
6 10 11 14 15   7 (a) N (b) A 20 N 20
8 (a) 9 (b) 21   9 −4
10 (a) points (8, 4) and (5, 1) plotted
  (b) three more points plotted, e.g. (6,2), (7,3), (4,0)

Pre-algebra test 9

1 (a) ☞ = 17 (b) ✳ = 38 (c) ✦ = 6½
2 (a) ☞ = 12 (b) ✳ = 12 (c) ✦ = 12
3 (a) ☞ = 157 (b) ✳ = 35 (c) ✦ = 48
4 (a) ☞ = 5 (b) ✳ = 7 (c) ✦ = 20
5 (a) ☞ = 7 (b) ✳ = 75 (c) ✦ = 20
6 (a) + (b) ÷ − (c) + ×
7 (a) − × (b) − × (c) + ÷ or − ×
8 (a) ÷ ÷ (b) ÷ − (c) + × +
9 (a) 20 cm (b) 10 cm 10 (a) 144 cm³ (b) 4 cm

Pre-algebra test 10

1 multiply by 5 and then divide by 8
2 (a) 108 (b) 2   3 +12 or ×1½

Mathematics Workbook: 10-minute Maths Tests Age 8–10 published by Galore Park

4 13
5 ♦ = 12 🐟 = 4
6 9 13 14 18
7 (a) L (b) Y
8 (a) 3 (b) −1 (c) $3\frac{1}{2}$
9 ÷ 2
10 (a) points (6, 3) and (8, 4) plotted
   (b) three more points plotted, e.g. (10, 5), (4, 2), (2, 1)

Shape, space and measures test 1

1 (a) 74 mm (b) 0.45 litres (c) 5.8 kg
2 (a) 1 kg (b) 40 cm
3 (a) A 5.75 B 2.2 (b) C 1.6 D 0.8
4 (a) 300 ml
   (b) 300 ml drawn in **B**
5 (a) 26 cm (b) 40 cm²
6 (a) 34 cm (b) 45 cm²
7 (a) 19:45 (b) 2.28 p.m. (c) Saturday
8 **A** isosceles right-angled triangle **B** parallelogram
   C square **D** rhombus
9 [shapes with lines of symmetry]
10 [arrow shape reflected in mirror line m]

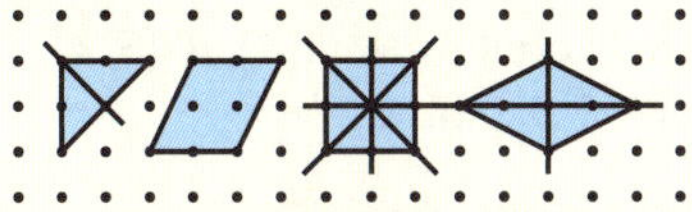

Shape, space and measures test 2

1 (a) regular hexagon (b) octagon
2 (a) 6 faces 12 edges 8 vertices
   (b) [net of cuboid]
3 (a) 2 cm³ (b) 10 cm²
4 a obtuse b right c acute d reflex
5 (a) 60° (b) 20°
6 (a) AD and BC (b) EF and HG (c) EH and FG
7 (a) angles A and C or B and D marked with stars
   (b) angles E and F or H and G marked with stars
8 (a) A, B and C plotted
   (b) triangle ABC drawn 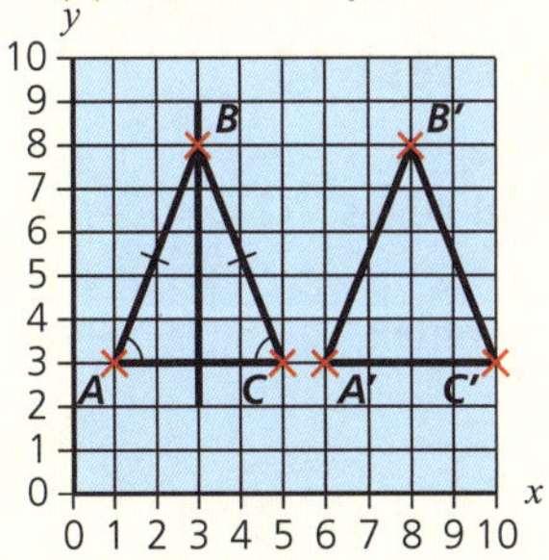
9 (a) isosceles triangle
   (b) AB and BC marked with dashes
   (c) angles A and C marked with arcs
   (d) line of symmetry x = 3 drawn
10 triangle ABC translated to A' (6, 3), B' (8, 8), C' (10, 3) A'B'C' plotted

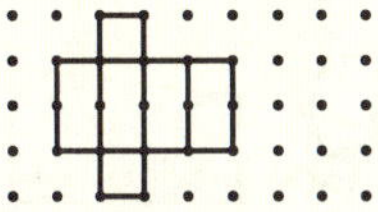

Shape, space and measures test 3

1 (a) 57 mm (b) 3000 ml (c) 35 000 g
2 (a) 1 g (b) 1 m
3 (a) A 8.5 B 0.7 (b) C −1 D 0.2
4 350 ml drawn in **A** and **B**
5 (a) 34 cm (b) 70 cm²
6 (a) 52 cm (b) 117 cm²
7 (a) 22:45 (b) 12.28 p.m. (c) Sunday
8 **A** rectangle **B** isosceles triangle **C** hexagon
   D parallelogram
9 [shapes with lines of symmetry: A, B, C, D]
10 [arrow shape reflected in mirror line m]

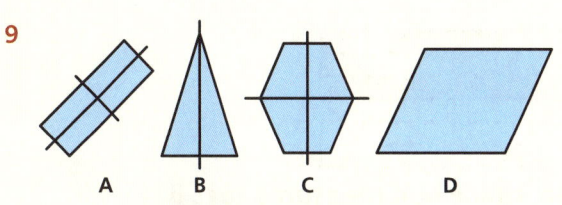

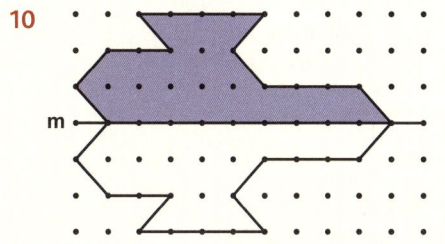

Shape, space and measures test 4

1 (a) regular pentagon (b) decagon
2 (a) 6 faces 12 edges 8 vertices
   (b) E.g. [net of cube]
3 (a) 8 cm³ (b) 24 cm²
4 a obtuse b right c acute d reflex
5 (a) 65° (b) 25°
6 (a) AD and BC (b) EF and HG (c) Any two sides
7 (a) angles A and C or B and D marked with stars
   (b) angles F and G marked with stars
8 (a) A, B, C and D plotted (b) kite ABCD drawn 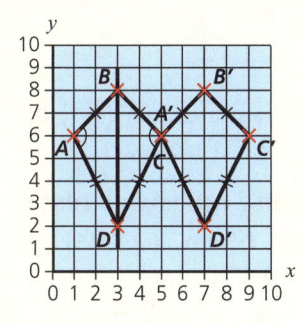
9 (a) kite
   (b) AB and BC or CD and DA marked with dashes
   (c) angles A and C marked with arcs
   (d) line of symmetry x = 3 drawn
10 kite ABCD translated to A' (5, 6), B' (7, 8), C' (9, 6), D' (7, 2) A'B'C'D' plotted

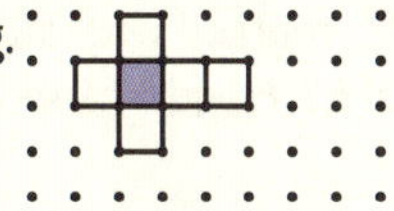

Shape, space and measures test 5

1 (a) 40 mm (b) 35 mm (c) 3000 mg
2 (a) 50 cm² (b) 250 ml
3 (a) A 13.5 B 1.6 (b) C 0.8 D −1
4 350 ml drawn in **B**
5 (a) 38 cm (b) 88 cm²
6 (a) 48 cm (b) 104 cm²
7 (a) 23:05 (b) 11.05 a.m. (c) Saturday
8 **A** isosceles trapezium **B** equilateral triangle
   C pentagon **D** rhombus
9 [shapes A, B, C, D]

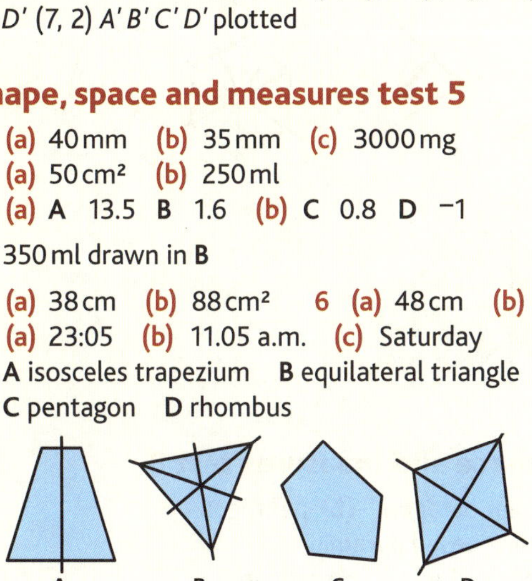

Mathematics Workbook: 10-minute Maths Tests Age 8–10 published by Galore Park

10 [shape reflected in mirror line m]

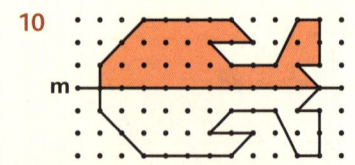

Shape, space and measures test 6

1 (a) regular decagon   (b) octagon
2 (a) 6 faces   12 edges   8 vertices
   (b) 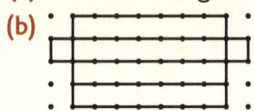 [net of cuboid]
3 (a) 7 cm³   (b) 30 cm²
4 a reflex   b obtuse   c right   d acute
5 e 39°   f 117°
6 (a) AB and DC
   (b) EF and IH or FG and IJ or GH and JE
   (c) EF and HI or any two of FG, GH, IJ and JE
7 (a) angles A and C or B and D marked with stars
   (b) angles G and J or any two of E, F, H and I marked with stars
8 (a) A, B, C and D plotted
   (b) square ABCD drawn

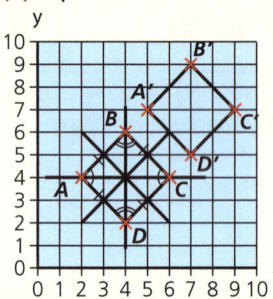

9 (a) square
   (b) any two sides marked with a dash
   (c) any two angles marked with an arc
   (d) four lines of symmetry drawn

10 square ABCD translated to A' (5, 7), B' (7, 9), C' (9, 7) and D' (7, 5) A'B'C'D' plotted

Shape, space and measures test 7

1 (a) 50 mm   (b) 0.3 cm   (c) 0.45 kg
2 (a) 100 cm   (b) 80 cm²
3 (a) A 3.5   B 1.3   (b) C −0.7   D 0.7
4 400 ml drawn in B
5 (a) 13 cm   (b) 10 cm²   6 (a) 52 cm   (b) 100 cm²
7 (a) 17:45   (b) 9.35 p.m.   (c) Sunday
8 A rectangle   B isosceles triangle
   C regular hexagon   D parallelogram
9 [shapes A, B, C, D]
10 [shape reflected in mirror line m]

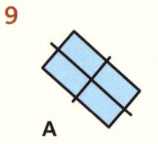

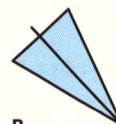

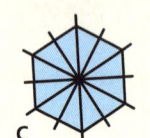

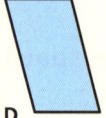

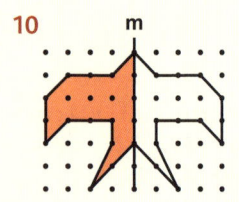

Shape, space and measures test 8

1 (a) regular octagon   (b) pentagon
2 (a) square-based pyramid
   (b) 5 faces   8 edges   5 vertices
3 (a) 60 cm³   (b) 94 cm²

4 a right   b reflex   c obtuse   d acute
5 (a) 46°   (b) 72°
6 (a) AD and BC   (b) EF and FG; GH and HE
7 (a) angles A and B or C and D marked with stars
   (b) angles E and G marked with stars
8 (a) A, B, C and D plotted
   (b) parallelogram ABCD drawn

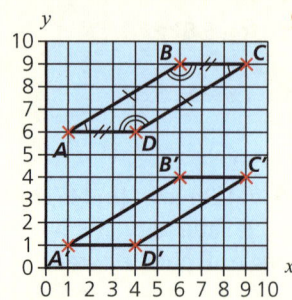

9 (a) parallelogram
   (b) sides AB and CD or BC and DA marked with dashes
   (c) angles A and C or B and D marked with arcs
   (d) no lines of symmetry

10 parallelogram ABCD translated to A' (1, 1), B' (6, 4), C' (9, 4) and D' (4, 1) A'B'C'D' plotted

Shape, space and measures test 9

1 128 mm   2 (a) 50 g   (b) 10 cm
3 (a) A 7.25   B 0.5   (b) C −0.5   D 3.7
4 (a) 850 ml   (b) 3 glasses, 190 ml left
5 18 cm²   6 (a) 16 cm   (b) 12 cm²
7 (a) 19:50   (b) 2.55 p.m.   (c) Sunday
8 A regular hexagon   B rectangle   C regular pentagon   D trapezium
9 [shapes A, B, C, D]
10 [shape reflected in mirror line m]

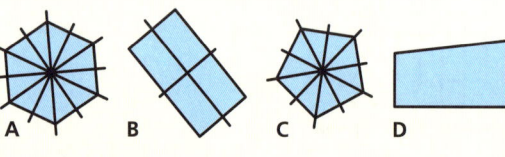

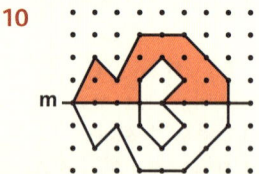

Shape, space and measures test 10

1 (a) regular heptagon   (b) decagon
2 (a) rectangles   (b) 8 faces   18 edges   12 vertices
3 (a) 10 cm³   (b) 33 cm²
4 a acute   b right   c obtuse   d reflex
5 (a) 41°   (b) 57°
6 (a) angles A and C or B and D, and F and G marked with stars
   (b) Any 2 of AB, CD, BC or DA, and EF and EG marked with dashes
7 (a) trapezium drawn
   (b) kite drawn
8 (a) A, B, C and D plotted
   (b) rhombus ABCD drawn 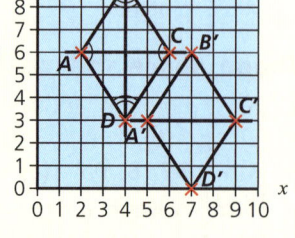
9 (a) rhombus
   (b) angles A and C or B and D marked with arcs
   (c) two lines of symmetry drawn

10 rhombus *ABCD* translated to *A'* (5, 3), *B'* (7, 6), *C'* (9, 3) and *D'* (7, 0) *A'B'C'D'* plotted

Handling data test 1
1 (a) 11   (b) 2
2 (a) Sofia   (b) Anne
3 Anne
4 (a) $\frac{1}{5}$ (b) 60%
5 (a) chocolate   (b) 4
6 25
7 (a) $\frac{1}{5}$ (b) 36%

8 / 9

Score	Tally of marks	Frequency
1	𝍸𝍸	10
2	𝍸 II	7
3	𝍸 III	8
4	𝍸𝍸	10
5	𝍸	5
6	𝍸𝍸	10

10 frequency chart plotted

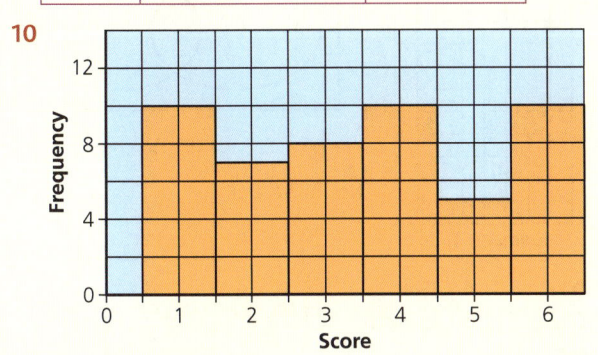

Handling data test 2
1 Venn diagram: Horizontal line of symmetry: B C D E; Vertical line of symmetry: A; intersection: H I; outside: F G J

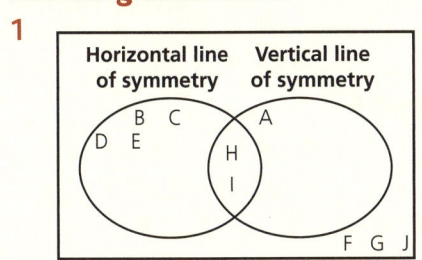

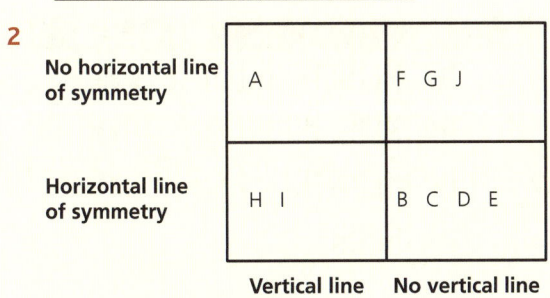

2 Carroll diagram:
- No horizontal line of symmetry / Vertical line of symmetry: A
- No horizontal line of symmetry / No vertical line of symmetry: F G J
- Horizontal line of symmetry / Vertical line of symmetry: H I
- Horizontal line of symmetry / No vertical line of symmetry: B C D E

3 (a) silver grey   (b) $\frac{1}{4}$
4 bar chart of colour vs number of cars
5 50%
6 (a) £1   (b) 20
7 (a) £5.50   (b) £12
8 getting 'heads' when a coin is tossed
9 (a) F   (b) A
10 (a) D   (b) D

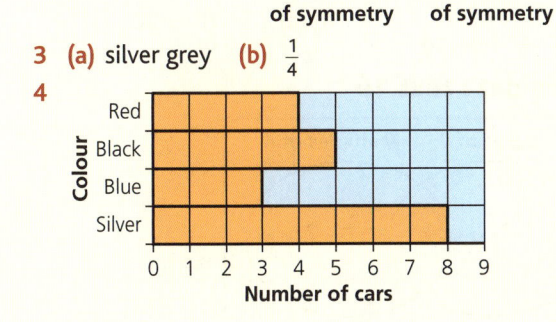

Handling data test 3
1 (a) Natalie   (b) 2
2 (a) Natalie   (b) Freya
3 31
4 (a) $\frac{2}{5}$ (b) 40%
5 (a) Spotty   (b) 3
6 20
7 (a) $\frac{1}{4}$ (b) 20%

8 / 9

Score	Tally of marks	Frequency
1	𝍸𝍸𝍸𝍸 III	23
2	𝍸𝍸𝍸 IIII	19
3	𝍸𝍸𝍸 III	18
4	𝍸𝍸𝍸𝍸	20
5	𝍸𝍸𝍸𝍸	20

10 frequency chart plotted

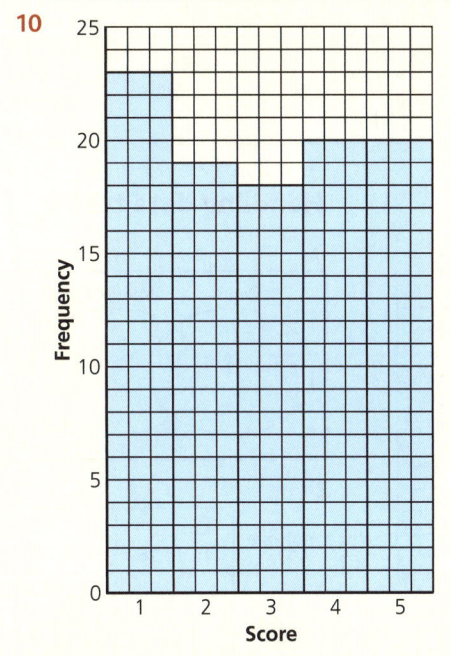

Handling data test 4
1 Venn diagram: Multiple of 2: 2 4 8; Multiple of 3: 3 9; intersection: 6; outside: 5 7 1

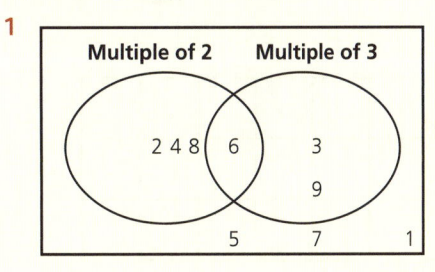

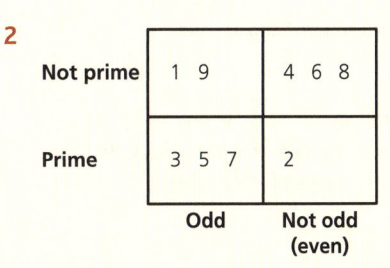

2 Carroll diagram:
- Not prime / Odd: 1 9
- Not prime / Not odd (even): 4 6 8
- Prime / Odd: 3 5 7
- Prime / Not odd (even): 2

3 (a) $\frac{1}{4}$ (b) $\frac{1}{3}$

Mathematics Workbook: 10-minute Maths Tests Age 8–10 published by Galore Park

4

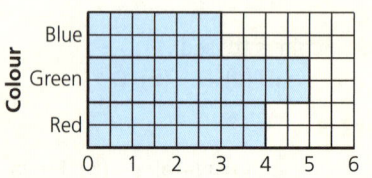

[bar chart: Blue 6, Green 4, Red 3 marbles]

5 40%
6 (a) 2 km (b) 1½ km
7 (a) 2 hours (b) 1 hour 45 minutes
8 scoring 1 when the spinner is spun
9 (a) B (b) A
10 (a) D (b) G

Handling data test 5

1 (a) Maria (b) 3
2 (a) Liam (b) Nicole
3 Maria
4 (a) 3/5 (b) 40%
5 (a) £9 (b) £3
6 £50
7 (a) 26% (b) £6.50
8 (a)

Score	Tally of marks	Frequency
1	卌 卌 II	12
2	卌 卌 I	11
3	卌 卌 卌	15
4	卌 卌 II	12

(b) 50

9 24%

10 [bar chart: Score 1→12, 2→11, 3→15, 4→12]

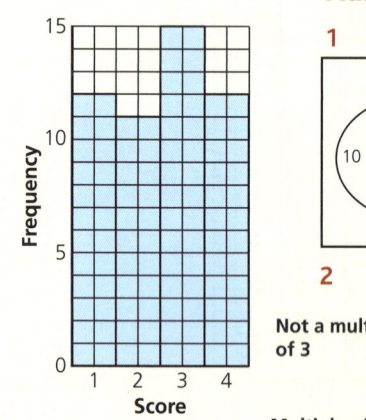

Handling data test 6

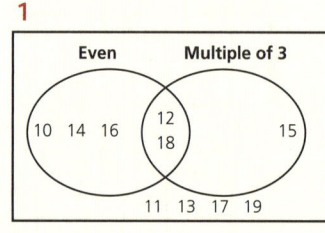

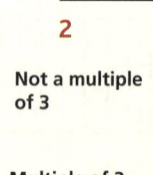

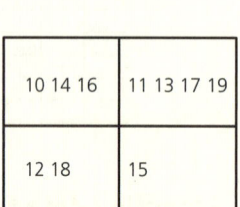

1 [Venn diagram: Even {10, 14, 16}, intersection {12, 18}, Multiple of 3 {15}, outside {11, 13, 17, 19}]

2

	Not a multiple of 3	Multiple of 3
Even	10 14 16	12 18
Not even	11 13 17 19	15

3 (a) 2 (b) 12

4 [bar chart: Milk 6, Dark 4, White 2 chocolates]

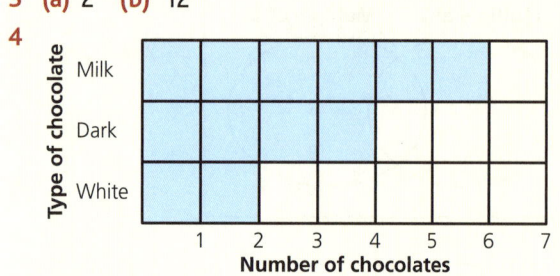

5 (a) ½ (b) 1/6
6 (a) 10 cm (b) 8 inches
7 (a) 30 cm (b) 180 cm
8 scoring a number less than 4 when an ordinary die is rolled
9 (a) G (b) A
10 (a) D (b) C

Handling data test 7

1 (a) Eric (b) Violet
2 (a) 18 (b) 2
3 2
4 (a) 1/3 (b) 50%
5 (a) 55 (b) 200
6 £240
7 (a) 1/8 (b) 25%
8 (a) 20
(b)

Score	Tally of marks	Frequency
'heads' (H)	卌 卌 II	
'tails' (T)	卌 III	

9 (a)

Score	Tally of marks	Frequency
'heads' (H)	卌 卌 II	12
'tails' (T)	卌 III	8

(b) 60%

10 [bar chart: Heads 12, Tails 8]

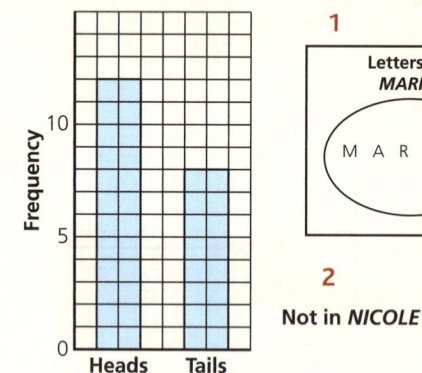

3 60%   **4** 25
5 60%   **6** 13.6 cm
7 Friday
8 getting blue when the spinner is spun
9 C

Handling data test 8

1 [Venn diagram: Letters in MARIE {M, A, R}, intersection {I, E}, Letters in NICOLE {N, C, O, L}, outside {D, F, H}]

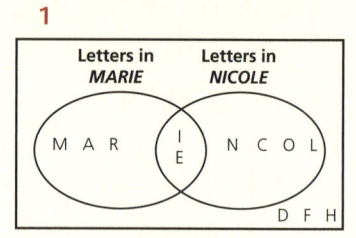

2

	Vowel	Not vowel
Not in NICOLE	A	D F H M R
In NICOLE	I O E	N C L

10 E and F

Handling data test 9

1 (a) Haley (b) 5
2 (a) Ben (b) Emma
3 Mia
4 (a) 2/7 (b) 60%
5 (a) 5 (b) 5
6 30
7 20%
8 (a)

Vehicle	Tally	Frequency
Car	卌 卌 卌 卌 卌 III	28
Van	卌 卌 卌 卌 IIII	24
HGV	卌 卌 卌 卌 I	21
Other	卌 卌 卌 II	17

(b) 90

9 360
10 (a) 53 (b) 300

Handling data test 10

1 [Venn diagram: Multiples of 3 {21, 27}, intersection {24}, Multiples of 4 {20, 28}, outside {22, 23, 25, 26, 29}]

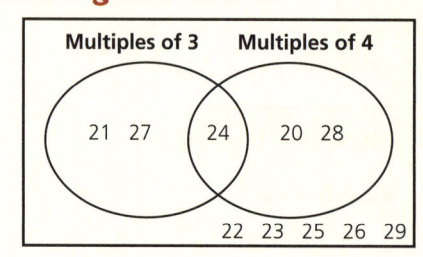

2

	Odd	Not odd
Not a multiple of 5	21 23 27 29	22 24 26 28
A multiple of 5	25	20

3 3

4 Bar chart: Lindsay, Jamie, Fido, Uneaten vs Number of slices (0–4)

5 $\frac{1}{4}$

6 Week 12 £70
7 45
8 seeing a man with a mass of more than 100 kg
9 A   **10** C

Mixed test 1

1 (a) grid 1–36 with 3, 5, 6, 12, 15, 21, 24, 25, 33, 35, 36 marked **(b)** 45

2 (a) 2105 **(b)** three thousand and fifty-six **(c)** 157
3 (a) 1450 **(b)** 97 100
4 (a) any half of shape shaded **(b)** $\frac{4}{12}$ **(c)** $\frac{1}{8} \frac{1}{3} \frac{1}{2}$ **(d)** $\frac{4}{3}$
5 (a) 5 **(b)** 2
6 (a) 1650 **(b)** 330 **(c)** 110
7 (a) 1081 **(b)** 37
8 (a) 2 hours 15 minutes **(b)** 2 lb 4 oz
9 (a) odd **(b)** odd odd
10 (a) $\frac{1}{4}$ **(b)** 15

Mixed test 2

1 Pattern 4 Pattern 5 (on grid)

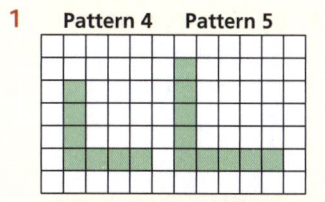

2

Pattern number	1	2	3	4	5
Area (square units)	1	3	5	7	9
Perimeter (units)	4	8	12	16	20

3 (a) 🐟 = 45 **(b)** ✽ = 55 **4 (a)** red **(b)** purple
5 three more points plotted, e.g. (1, 2), (2, 3), (4, 5)
6 A 0.7 B 2.35

7 (shape on dot grid with mirror line m)

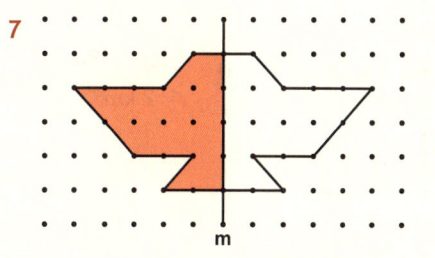

8 (a) 62° **(b)** 18°
9 £25.20
10 scoring 3 or more with the die

Mixed test 3

1 (a) 6
  (b) 4 5 8 10 20
  (c) 1020 920 820 720
2 (a) 2007
  (b) three hundred and five thousand and nineteen
  (c) 7000
3 (a) 51 500 **(b)** 12 **(c)** 5.4
4 (a) 0.89 **(b)** $\frac{4}{5}$ **(c)** $\frac{3}{10}$ **(d)** 0.28

5

×	9	6	8	7
6	54	36	48	42
8	72	48	64	56
7	63	42	56	49
9	81	54	72	63

6 (a) 401 **(b)** 3 **(c)** £3.45
7 (a) 2820 **(b)** 502
8 (a) 5 hours 24 minutes **(b)** 5 cm 4 mm
9 (a) 2 **(b)** 12 000   **10 (a)** 29 **(b)** 61

Mixed test 4

1 (a) 47 → 15 → 7 **(b)** 99 → 27 → 11 → 3
2 65
3 (a) 🐟 = 13 **(b)** ✽ = 75   **4 (a)** A **(b)** B
5 (a) +2 **(b)** 10
6 800 ml drawn in A
7 (shape reflected across line m on dot grid)

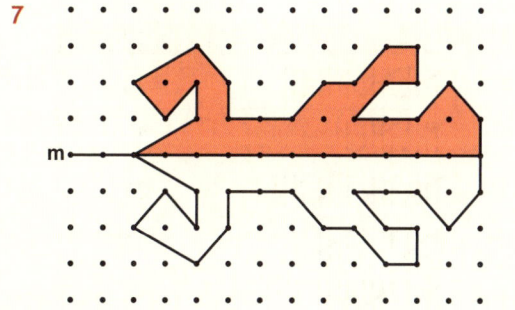

8 (a) points plotted **(b)** parallelogram
9 (a) Lara **(b)** Lara 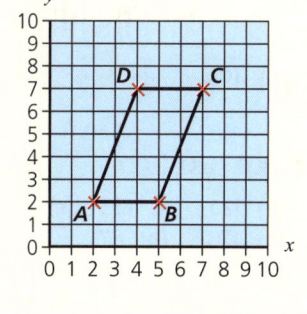
10 even chance

Parallelogram ABCD plotted on grid: A(4,2), B(7,2), C(8,7), D(5,7)

Mixed test 5

1 (a) 13 23 43 53
    73 circled
  (b) 4 °C
  (c) 64

2 (a) 40000
  (b) M HTh TTh Th H T U — 9 1 0 0 0 2
  (c) 3080
3 (a) number line showing −3, 0, 5, 6.5, 10
  (b) 400   (c) 9.7
4 (a) any quarter of shape shaded
  (b) $\frac{12}{20}$ (c) $\frac{3}{7}$ $\frac{3}{5}$ $\frac{3}{4}$
5 (a) 43 − 14 = 29
      29 + 14 = 43
      14 + 29 = 43
  (b) 156 ÷ 13 = 12
      12 × 13 = 156
      13 × 12 = 156
6 (a) 840   (b) 1728
7 (a) 5 remainder 5   (b) 15
8 (a) 2 hours 45 minutes   (b) 2 feet 9 inches
9 (a) even   (b) even
10 (a) £8   (b) £2   £1

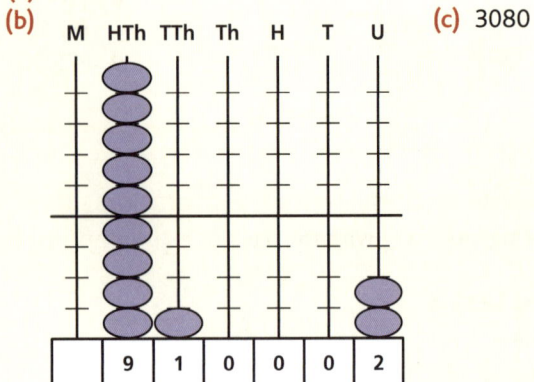

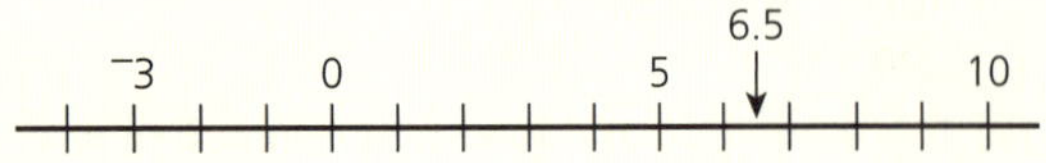

Mixed test 6
1 Pattern 4
2 14
3 (a) +   (b) ÷   −
4 (a) E   (b) 20
5 11
6 (a) 45 cm²   (b) 28 cm
7 A parallelogram   B regular hexagon   C rhombus   D isosceles trapezium
8 (a) a reflex   b obtuse   c acute
  (b) d = 43°   e = 128°
9 (a)

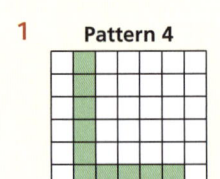

Score	Tally of scores	Frequency
1	⊞⊞III	18
2	⊞⊞II	12
3	⊞⊞⊞⊞I	21
4	⊞⊞⊞II	17
5	⊞⊞⊞	15
6	⊞⊞⊞II	17

  (b) 49%
10 D

Mixed test 7
1 (a) 121   (b) 125   (c) 42 45 48
2 (a) 4763   (b) 57
3 (a) number line 0 – 4 – 10
  (b) 380000   (c) 9.0
4 (a) $\frac{4}{5}$ (b) $\frac{1}{4}$ (c) 40
5
×	5	9	7	6
7	35	63	49	42
9	45	81	63	54
8	40	72	56	48
6	30	54	42	36

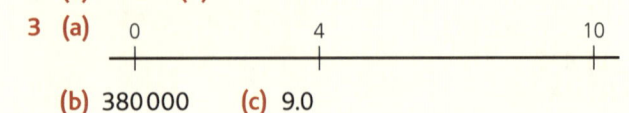

6 (a) 110   (b) 496
7 (a) 1081   (b) 560
8 (a) 3   (b) 40 cm
9 (a) 0   (b) 16 000
10
	A	B
Peely	07:59	11:15
Quewty	09:30	12:46
Ruffly	11:52	15:08

Mixed test 8
1 (a) 39 → 21 → 4
  (b) 98 → 25 → 12 → 5
  (c) 49 → 22 → 8
2 11, 15, 17, 18
3 (a) ✻ = 24   (b) ✦ = 22
4 N
5 (3, 1) and three more points plotted
6 A 9.25   B 2.2
7 (four shapes with lines of symmetry shown)
8 (a) coordinate grid with points A, B, C, D and A', B', C', D'
  (b) shape translated, with points at A' (2, 2), B' (5, 4), C' (8, 2) and D' (5, 0)
9 (a) Niamh   (b) Kiera
10 certain

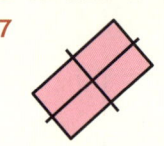

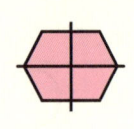

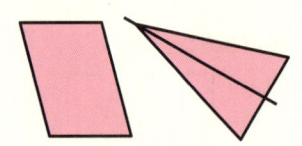

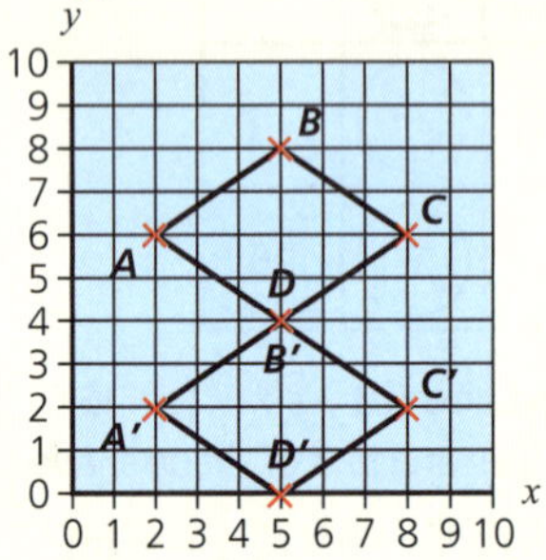

Mixed test 9
1 105
2 (a) 2878   (b) 4.53   (c) 79
3 (a) 206 000   (b) 2 000 000   (c) 14   (d) 18.0
4 (a) $\frac{2}{3}$ (b) $1\frac{1}{3}$ (c) 36
5
×	6	12	7	9
8	48	96	56	72
7	42	84	49	63
9	54	108	63	81
11	66	132	77	99

6 (a) 30   (b) £8.11
7 (a) 3804   (b) 143
8 (a) 2 hours 3 minutes   (b) 2 litres 50 ml
9 (a) even   (b) odd
10 $\frac{2}{3}$

Mixed test 10
1
Pattern number	1	2	3	4	5
Number of small triangles	1	4	9	16	25
Perimeter (units)	3	6	9	12	15

2 64
3 (a) ÷   (b) −   ×
4 (a) A   (b) 30
5 $\frac{1}{2}$
6 (a) 30 cm²   (b) 23 cm
7 A octagon   B isosceles right-angled triangle   C parallelogram   D equilateral triangle
8 (a) a obtuse   b acute   c reflex   (b) d = 41°   e = 64°
9 56
10 B

A12 Mathematics Workbook: 10-minute Maths Tests Age 8–10 published by Galore Park

Pre-algebra test 3

Final score: 20 ___ %

Questions 1 to 5 concern missing numbers.

1. Find the number represented by the symbol in each of these sentences.
   - (a) 9 + ◆ = 25   ◆ = _____
   - (b) ✸ − 11 = 12   ✸ = _____
   - (c) ◆ + ◆ = 26   ◆ = _____

2. Find the number represented by the symbol in each of these sentences.
   - (a) 7 × ◆ = 56   ◆ = _____
   - (b) ✸ × 9 = 81   ✸ = _____
   - (c) ◆ × ◆ = 49   ◆ = _____

3. Find the number represented by the symbol in each of these sentences.
   - (a) 117 + ◆ = 132   ◆ = _____
   - (b) ✸ − 33 = 29   ✸ = _____
   - (c) 71 − ◆ = 24   ◆ = _____

4. Find the number represented by the symbol in each of these sentences.
   - (a) 23 × ◆ = 69   ◆ = _____
   - (b) ✸ + 19 = 42   ✸ = _____
   - (c) 44 − ◆ = ◆   ◆ = _____

5. Find the number represented by the symbol in each of these sentences.
   - (a) 12 + ◆ = 36 − ◆   ◆ = _____
   - (b) ✸ ÷ 6 = 12   ✸ = _____
   - (c) $\frac{1}{3}$ ◆ = 24   ◆ = _____

Questions 6 to 8 concern missing operation signs.

6. On each line write the operation sign (− + × ÷) to make the sentences true.
   - (a) 8 _____ 14 = 22
   - (b) 36 _____ 6 = 5 _____ 6
   - (c) 9 _____ 3 _____ 7 = 5

7. In each of these, write the operation signs (− + × ÷) on the lines, to make the sentences true.
   - (a) 7 _____ 11 = 36 _____ 2
   - (b) 9 _____ 2 = 6 _____ 3
   - (c) (9 _____ 2) _____ 5 = 35

8. In each of these, write the operation signs (− + × ÷) on the lines, to make the sentences true.
   - (a) 5 _____ 4 _____ 2 = 40
   - (b) 8 _____ 5 = 17 _____ 4
   - (c) (6 _____ 2) _____ (9 _____ 3) = 24

Questions 9 and 10 concern simple word formulae.

9. The perimeter of a regular hexagon can be found using the word formula:

   Multiply the length of one side by six.

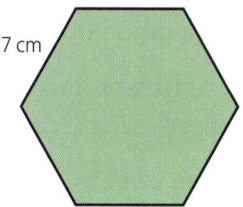

   7 cm

   - (a) What is the perimeter of the regular hexagon above? _____ cm
   - (b) What would be the length of a side of a regular hexagon with perimeter 30 cm? _____ cm

10. The volume of a cuboid can be found using the word formula:

    Multiply the length by the width by the height.

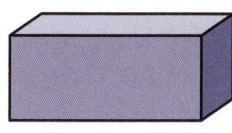

    - (a) What is the volume of a cuboid which is 6 cm long, 2 cm wide and 3 cm high? _____ cm³
    - (b) What would be the height of a cuboid which is 10 cm long, 5 cm wide and has volume 150 cm³? _____ cm

39

Pre-algebra test 4

Final score
20 ___ %

1. Masses in grams can be converted into masses in kilograms by using the simple word formula:

   Divide by a thousand.

   Write a similar word formula which can be used to convert masses in kilograms into masses in grams. _____

Questions 2 to 5 concern number puzzles.

2. (a) Sacha was asked to think of a number and she chose 13

   She was asked to multiply her number by 3 and write the result.

   What did she write? _____

   (b) Jasmine was asked to think of a number and perform the operation ÷ 5

   She wrote down 12

   What number did she think of? _____

3. Sean was given the number 21 and asked to perform an operation on it. 7 was the result.

   Suggest two different operations that Sean might have done. _____ or _____

4. Jamie has some sweets in his pocket but he does not know how many there are.

   He takes 10 sweets from his pocket to give 5 each to Lucy and Emma. Emma gives Jamie 7 sweets and Lucy gives him 6 sweets. Jamie eats 3 sweets and when he gets home he finds that he has 27 sweets.

   How many sweets did Jamie have at the start? _____

5. In this question ♦ and ☛ are two integers (whole numbers).

   ♦ − ☛ = 12 and ♦ + ☛ = 24

   What are ♦ and ☛? ♦ = _____ and ☛ = _____

Questions 6 and 7 concern sequences.

6. (a) Draw the next three beads in this sequence.

   ○●●○○●●○○○●○○

   (b) What is the colour of the next bead in this sequence? _____

   ●●○●●○●●○●●○●

7. Study this sequence of alphabet letters.
   ABACADAEA

   The 1st letter is **A** and the 6th letter is **D**. What will be

   (a) the 11th letter _____

   (b) the 14th letter _____

   (c) the 49th letter? _____

Questions 8 to 10 concern number machines.

8. Chris has a machine which multiplies any number that he puts into it by 4

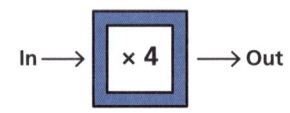

   In ⟶ [× 4] ⟶ Out

   (a) If Chris puts in 6, what number comes out? _____

   (b) If 60 comes out, what number did Chris put in? _____

9. Craig has a machine, but the label has fallen off and he can't remember what it does!

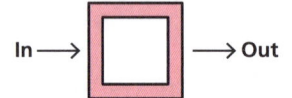

   In ⟶ [ ] ⟶ Out

   If Craig puts in 2, 6 comes out. If he puts in 3, 7 comes out.

   Write the operation on the label.

10. (a) On the grid, mark with crosses the points which represent the (in, out) numbers (2, 6) and (3, 7) for Craig's machine.

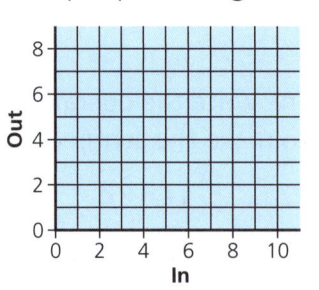

    (b) Mark with crosses three more (in, out) pairs of numbers for Craig's machine.

40

Pre-algebra test 5

Final score
20 ___ %

Questions 1 to 5 concern missing numbers.

1. Find the number represented by the symbol in each of these sentences.

    (a) $13 + \bullet = 31$ $\bullet = $ _____

    (b) $\ast - 9 = 17$ $\ast = $ _____

    (c) $\diamond + \diamond = 42$ $\diamond = $ _____

2. Find the number represented by the symbol in each of these sentences.

    (a) $8 \times \bullet = 48$ $\bullet = $ _____

    (b) $\ast \times 7 = 63$ $\ast = $ _____

    (c) $\diamond \times \diamond = 81$ $\diamond = $ _____

3. Find the number represented by the symbol in each of these sentences.

    (a) $103 + \bullet = 121$ $\bullet = $ _____

    (b) $\ast - 19 = 35$ $\ast = $ _____

    (c) $55 - \diamond = 38$ $\diamond = $ _____

4. Find the number represented by the symbol in each of these sentences.

    (a) $14 \times \bullet = 42$ $\bullet = $ _____

    (b) $\ast + 37 = 71$ $\ast = $ _____

    (c) $50 - \diamond = \diamond$ $\diamond = $ _____

5. Find the number represented by the symbol in each of these sentences.

    (a) $8 + \bullet = 32 - \bullet$ $\bullet = $ _____

    (b) $\ast \div 4 = 16$ $\ast = $ _____

    (c) $\frac{1}{4} \diamond = 4$ $\diamond = $ _____

Questions 6 to 8 concern missing operation signs.

6. In each of these, write the operation signs ($- + \times \div$) on the lines, to make the sentences true.

    (a) 35 _____ $17 = 18$

    (b) 20 _____ $5 = 3$ _____ 1

    (c) 7 _____ 3 _____ $6 = 16$

7. In each of these, write the operation signs ($- + \times \div$) on the lines, to make the sentences true.

    (a) 13 _____ $11 = 36$ _____ 12

    (b) 18 _____ $2 = 6$ _____ 3

    (c) $(4$ _____ $3)$ _____ $8 = 56$

8. In each of these, write the operation signs ($- + \times \div$) on the lines, to make the sentences true.

    (a) 2 _____ 3 _____ $4 = 24$

    (b) 9 _____ $3 = 12$ _____ 4

    (c) $(6$ _____ $3)$ _____ $(4$ _____ $1) = 27$

Questions 9 and 10 concern simple word formulae.

9. The perimeter of a regular pentagon can be found using the word formula:

    Multiply the length of one side by five.

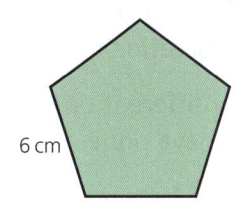

    6 cm

    (a) What is the perimeter of the regular pentagon above? _____ cm

    (b) What would be the length of a side of a regular pentagon with perimeter 70 cm? _____ cm

10. The volume of a cuboid can be found using the word formula:

    Multiply the length by the width by the height.

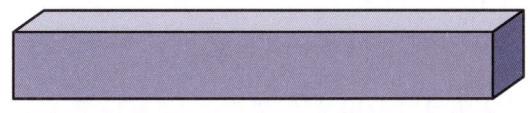

    (a) What is the volume of a cuboid which is 15 cm long, 2 cm wide and 2 cm high? _____ cm³

    (b) What would be the height of a cuboid which is 12 cm long, 3 cm wide and has volume 180 cm³? _____ cm

41

Pre-algebra test 6

Final score

20 ___ %

1. Lengths in feet can be converted into lengths in centimetres by using the simple word formula:

   Multiply by thirty.

   Write a similar word formula which can be used to convert lengths in centimetres into lengths in feet. _____

Questions 2 to 5 concern number puzzles.

2. (a) Rachel was asked to think of a number and she chose 7

   She was asked to multiply her number by 8 and write the result.

   What did she write? _____

   (b) Philippa was asked to think of a number and perform the operation ÷ 8

   She wrote down 4

   What number did she think of? _____

3. Yusuf was given the number 24 and asked to perform an operation on it. The result was 8

   Suggest two different operations that Yusuf might have done. ____ or ____

4. Chloe has 12 £1 coins in her purse. She takes 4 coins to pay for a book priced at £4

   On the way out of the shop she drops her purse and the coins roll all over the path. She picks up all she can see and puts them back into her purse. When Chloe gets home she counts the money and finds she has 6 £1 coins.

   How many coins has she lost? _____

5. In this question ❖ and ☛ are two integers (whole numbers).

   ❖ − ☛ = 10 and ❖ + ☛ = 30

   What are ❖ and ☛? ❖ = ____ and ☛ = ____

Questions 6 and 7 concern sequences.

Eve is making a necklace of beads with the letters of her name. The beads in places 2, 5 and 8 are V.

E V E E V E E V

6. What will be the places of the next four V beads? _____ _____ _____ _____

7. Eve's completed necklace has a total of 30 beads.

   (a) What letter will the last bead be? _____

   (b) How many beads of each letter will there be?   E ____   V ____

Questions 8 to 10 concern number machines.

8. Ahmed has a machine which multiplies any number that he puts into it by 5

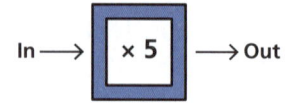

   In ⟶ [× 5] ⟶ Out

   (a) If Ahmed puts in 8, what number comes out? _____

   (b) If 80 comes out, what number did Ahmed put in? _____

9. Peter has a machine, but the label has fallen off and he can't remember what it does!

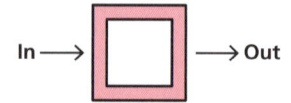

   In ⟶ [ ] ⟶ Out

   If Peter puts in 1, 4 comes out. If he puts in 3, 6 comes out. Write the operation on the label.

10. (a) On the grid, mark with crosses the points which represent the (in, out) numbers (1, 4) and (3, 6) for Peter's machine.

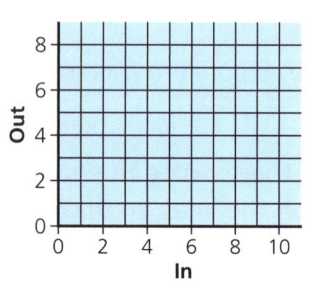

    (b) Mark with crosses three more points representing (in, out) pairs of numbers for Peter's machine.

42

Pre-algebra test 7

Final score: 20 ___ %

Questions 1 to 5 concern missing numbers.

1. Find the number represented by the symbol in each of these sentences.

    (a) 33 + ☛ = 52   ☛ = _____
    (b) ✻ − 12 = 24   ✻ = _____
    (c) ❖ + ❖ = 5   ❖ = _____

2. Find the number represented by the symbol in each of these sentences.

    (a) 7 × ☛ = 63   ☛ = _____
    (b) ✻ × 8 = 64   ✻ = _____
    (c) ❖ × ❖ = 121   ❖ = _____

3. Find the number represented by the symbol in each of these sentences.

    (a) 789 + ☛ = 1000   ☛ = _____
    (b) ✻ − 43 = 17   ✻ = _____
    (c) 72 − ❖ = 49   ❖ = _____

4. Find the number represented by the symbol in each of these sentences.

    (a) 15 × ☛ = 75   ☛ = _____
    (b) ✻ + 18 = 45   ✻ = _____
    (c) 70 − ❖ = ❖   ❖ = _____

5. Find the number represented by the symbol in each of these sentences.

    (a) 17 + ☛ = 23 − ☛   ☛ = _____
    (b) ✻ ÷ 3 = 21   ✻ = _____
    (c) $\frac{3}{4}$ ❖ = 6   ❖ = _____

Questions 6 to 8 concern missing operation signs.

6. In each of these, write the operation signs (− + × ÷) on the lines, to make the sentences true.

    (a) 45 _____ 9 = 54
    (b) 32 _____ 8 = 8 _____ 4
    (c) 5 _____ 4 _____ 3 = 60

7. In each of these, write the operation signs (− + × ÷) on the lines, to make the sentences true.

    (a) 25 _____ 31 = 7 _____ 8
    (b) 20 _____ 4 = 4 _____ 4
    (c) (3 _____ 3) _____ 2 = 12

8. In each of these, write the operation signs (− + × ÷) on the lines, to make the sentences true.

    (a) 6 _____ 2 _____ 3 = 4
    (b) 18 _____ 3 = 24 _____ 4
    (c) (9 _____ 3) _____ (5 _____ 1) = 2

Questions 9 and 10 concern simple word formulae.

9. The perimeter of a rhombus can be found using the word formula:

    Multiply the length of one side by four.

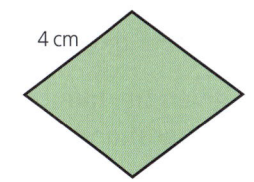

    4 cm

    (a) What is the perimeter of the rhombus above? _____ cm
    (b) What would be the length of a side of a rhombus with perimeter 28 cm? _____ cm

10. The volume of a cuboid can be found using the word formula:

    Multiply the length by the width by the height.

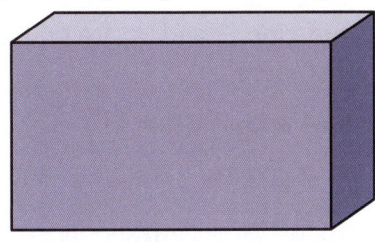

    (a) What is the volume of a cuboid which is 11 cm long, 5 cm wide and 6 cm high? _____ cm³
    (b) What would be the height of a cuboid which is 10 cm long, 4 cm wide and has volume 320 cm³? _____ cm

Pre-algebra test 8

Final score: 20 ___ %

1. Areas in square centimetres can be converted into areas in square millimetres by using the simple word formula:

    Multiply by one hundred.

    Write a similar word formula which can be used to convert areas in square millimetres into areas in square centimetres. _____

Questions 2 to 5 concern number puzzles.

2. (a) Tareq was asked to think of a number and he chose his lucky number 13

    He was asked to multiply his number by 5 and write the result.

    What did he write? _____

    (b) Serena was asked to think of a number and perform the operation ÷ 9

    She wrote down 12

    What number did she think of? _____

3. Clare was given the number 32 and asked to perform an operation on it. The result was 16

    Suggest two different operations that Clare might have done. _____ or _____

4. A coach driver set off from Ayby but forgot to count the number of passengers. At Beeford, 7 passengers got off and 13 got on. The 52-seater coach was then full.

    How many people were on the coach when it left Ayby? _____

5. In this question ❖ and ☛ are two integers (whole numbers).

    ❖ × ☛ = 24 and ❖ + ☛ = 11

    What are ❖ and ☛? ❖ = ____ and ☛ = ____

Questions 6 and 7 concern sequences.

Anna is making a necklace of beads with the letters of her name. The 2nd and 3rd beads are **N**.

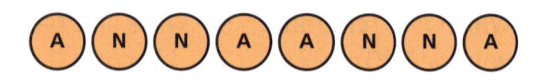

(A) (N) (N) (A) (A) (N) (N) (A)

6. What will be the places of the next four **N** beads? _____ _____ _____ _____

7. Anna's completed necklace has a total of 40 beads.

    (a) What letter will the 39th bead be? _____

    (b) How many beads of each letter will there be? A _____ N _____

Questions 8 to 10 concern number machines.

8. Alice has a machine which subtracts 4 from any number that she puts into it.

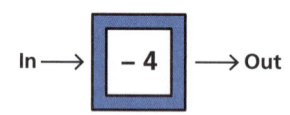

    In ⟶ [− 4] ⟶ Out

    (a) If Alice puts in 13, what number comes out? _____

    (b) If 17 comes out, what number did Alice put in? _____

9. Omar has a machine, but the label has fallen off and he can't remember what it does!

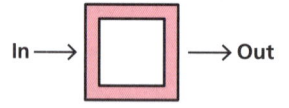

    In ⟶ [ ] ⟶ Out

    If Omar puts in 8, 4 comes out. If he puts in 5, 1 comes out.

    Write the operation on the label.

10. (a) On the grid, mark with crosses the points which represent the (in, out) numbers (8, 4) and (5, 1) for Omar's machine.

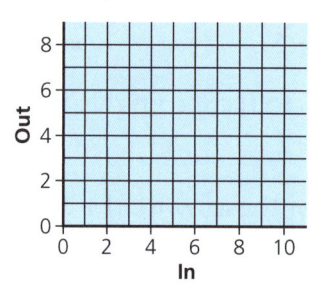

    (b) Mark with crosses three more points representing (in, out) pairs of numbers for Omar's machine.

44

Pre-algebra test 9

Final score

20 ___ %

Questions 1 to 5 concern missing numbers.

1. Find the number represented by the symbol in each of these sentences.
   (a) 86 + ☛ = 103   ☛ = _____
   (b) ✻ − 31 = 7   ✻ = _____
   (c) ❖ + ❖ = 13   ❖ = _____

2. Find the number represented by the symbol in each of these sentences.
   (a) 11 × ☛ = 132   ☛ = _____
   (b) ✻ × 6 = 72   ✻ = _____
   (c) ❖ × ❖ = 144   ❖ = _____

3. Find the number represented by the symbol in each of these sentences.
   (a) 343 + ☛ = 500   ☛ = _____
   (b) ✻ − 7 = 28   ✻ = _____
   (c) 53 − ❖ = 5   ❖ = _____

4. Find the number represented by the symbol in each of these sentences.
   (a) 16 × ☛ = 80   ☛ = _____
   (b) ✻ + 7 = 14   ✻ = _____
   (c) 40 − ❖ = ❖   ❖ = _____

5. Find the number represented by the symbol in each of these sentences.
   (a) 23 + ☛ = 37 − ☛   ☛ = _____
   (b) ✻ ÷ 5 = 15   ✻ = _____
   (c) $\frac{3}{5}$ ❖ = 12   ❖ = _____

Questions 6 to 8 concern missing operation signs.

6. In each of these, write the operation signs (− + × ÷) on the lines, to make the sentences true.
   (a) 43 _____ 17 = 60
   (b) 40 _____ 8 = 8 _____ 3
   (c) 1 _____ 2 _____ 3 = 7

7. In each of these, write the operation signs (− + × ÷) on the lines, to make the sentences true.
   (a) 30 _____ 6 = 6 _____ 4
   (b) 30 _____ 6 = 4 _____ 6
   (c) (5 _____ 4) _____ 3 = 3

8. In each of these, write the operation signs (− + × ÷) on the lines, to make the sentences true.
   (a) 8 _____ 2 _____ 2 = 2
   (b) 24 _____ 6 = 8 _____ 4
   (c) (3 _____ 5) _____ (3 _____ 4) = 56

Questions 9 and 10 concern simple word formulae.

9. The perimeter of a rectangle can be found using the word formula:

   Add the length to the width and then multiply by two.

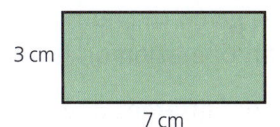

   3 cm
   7 cm

   (a) What is the perimeter of the rectangle above? _____ cm
   (b) What would be the length of a side of a rectangle which is 5 cm wide and has perimeter 30 cm? _____ cm

10. The volume of a cuboid can be found using the word formula:

    Multiply the length by the width by the height.

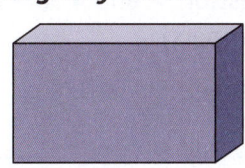

    (a) What is the volume of a cuboid which is 12 cm long, 4 cm wide and 3 cm high? _____ cm³
    (b) What would be the height of a cuboid which is 5 cm long, 3 cm wide and has volume 60 cm³ _____ cm

Pre-algebra test 10

Final score 20 ___ %

1. Distances in miles can be converted to distances in kilometres by using the word formula:

   Multiply by 8 and then divide by 5.

   Write a similar word formula which can be used to convert distances in kilometres to distances in miles. _____

   _____

Questions 2 to 5 concern number puzzles.

2. (a) Connor was asked to think of a number and he chose 9

   He was asked to multiply his number by 12 and write the result.

   What did he write? _____

   (b) Isabel was asked to think of a number and perform the operation ÷ 4

   She wrote down $\frac{1}{2}$.

   What number did she think of? _____

3. Oliver was given the number 24 and asked to perform an operation on it. The result was 36.

   Suggest two different operations that Oliver might have done. _____ or _____

4. A theatre can seat 240 people and there are just 11 seats left for a concert by the group No Direction. 6 people return tickets because of illness and 4 seats are booked at the last minute.

   How many empty seats are there when the concert starts? _____

5. In this question ❖ and ☛ are two integers (whole numbers).

   ❖ + ☛ = 16 and ❖ ÷ ☛ = 3

   What are ❖ and ☛? ❖ = _____ and ☛ = _____

Questions 6 and 7 concern sequences.

Billy is making a string of beads with the letters of his name. The 3rd and 4th beads are **L**.

B I L L Y B I L

6. What will be the places of the next four **L** beads? _____ _____ _____ _____

7. Billy completed string has a total of 60 beads.

   (a) What letter will the 59th bead be? _____

   (b) What letter will the 45th bead be? _____

Questions 8 to 10 concern number machines.

8. Alex has a machine which subtracts 3 from any number that he puts into it.

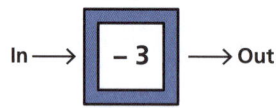

   In ⟶ [− 3] ⟶ Out

   (a) If 0 comes out, what number did Alex put in? _____

   (b) If Alex puts in 2, what number comes out? _____

   (c) If $\frac{1}{2}$ comes out, what number did Alex put in? _____

9. Chloe has a machine, but the label has fallen off and she can't remember what it does!

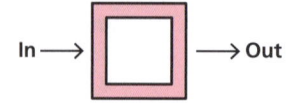

   In ⟶ [ ] ⟶ Out

   If Chloe puts in 6, 3 comes out. If she puts in 8, 4 comes out.

   Write the operation on the label.

10. (a) On the grid, mark with crosses the points which represent the (in, out) numbers (6, 3) and (8, 4) for Chloe's machine.

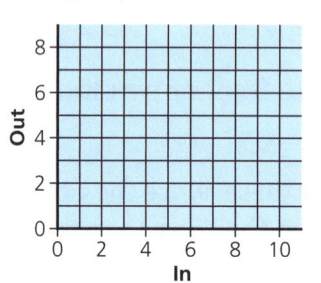

    (b) Mark with crosses three more points representing (in, out) pairs of numbers for Chloe's machine.

Shape, space and measures test 1

Final score
20 ___ %

1 **(a)** Measure the length of line *AB*, giving your answer to the nearest millimetre.
   _____ mm
   A _____ B

   (b) Write 450 ml in litres. _____ litres

   (c) Write 5800 g in kilograms. _____ kg

2 **(a)** Which of the following is the best estimate of the mass of a litre bottle of lemonade? _____

   10 g   100 g   500 g   1 kg   5 kg

   (b) Which of the following is the best estimate of the perimeter of a £5 note? _____

   10 cm   20 cm   40 cm   200 cm

3 **(a)** Write down the readings on scales **A** & **B**.
   A _____   B _____

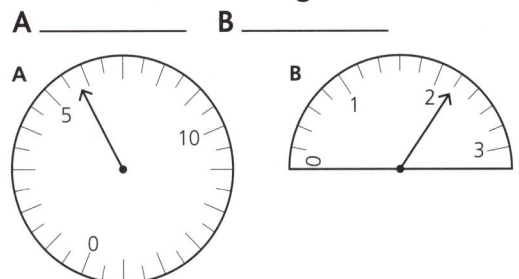

   (b) Write down the readings on scales **C** & **D**.
   C _____   D _____

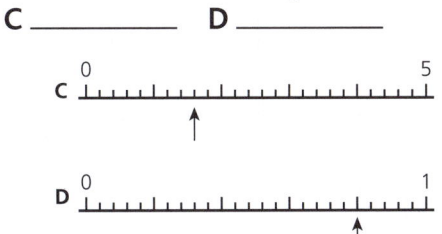

4 **(a)** What volume of liquid is in cylinder **A**?
   _____ ml

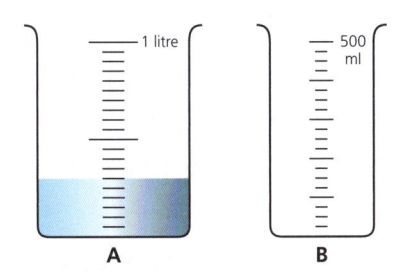

   (b) All of the liquid is poured from **A** into **B**.
   Draw the liquid in **B**.

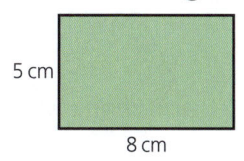

5 The diagram shows a rectangle.

   5 cm
   8 cm

   Calculate
   (a) the perimeter _____ cm
   (b) the area. _____ cm²

6 Look at the diagram.

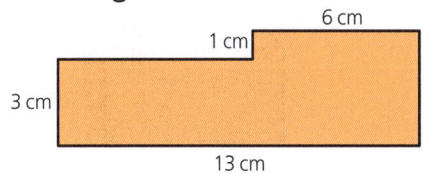

   6 cm
   1 cm
   3 cm
   13 cm

   Calculate
   (a) the perimeter _____ cm
   (b) the area. _____ cm²

7 **(a)** Write 7.45 p.m. as a 24-hour time.
   ____ : ____

   (b) Write 14:28 as a 12-hour time.
   _____

   (c) Amy's birthday is 5 October and her brother Tim's birthday is 24 September. This year Tim's birthday is on a Tuesday. On which day of the week will Amy's birthday be this year? _____

8 Name the plane shapes below.
   A _____   B _____
   C _____   D _____

   A   B   C   D

9 Draw all the lines of symmetry on the shapes **A**, **B**, **C** and **D** in question 8

10 Complete the shape below which is symmetrical about the line **m**.

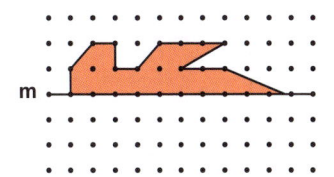

   m

47

Shape, space and measures test 2

Final score
20 ___ %

1 (a) Name this polygon. _____

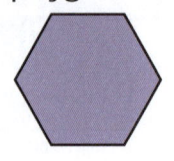

   (b) What is the name of a polygon with 8 sides? _____

2 The cuboid shown here is 2 cm long, 1 cm wide and 1 cm high.

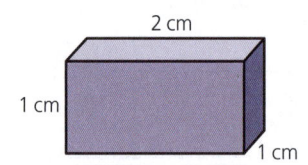

   (a) Complete: A cuboid has _____ faces, _____ edges and _____ vertices.

   (b) On the square dotted grid draw a net for the cuboid. One face is drawn for you.

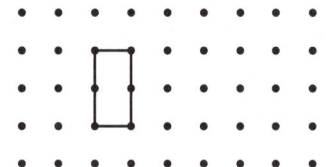

3 For the cuboid in question 2, what is

   (a) the volume of the cuboid _____ cm³
   (b) the area of card needed to make the cuboid? _____ cm²

4 Name the types of these angles. a _____
   b _____ c _____ d _____

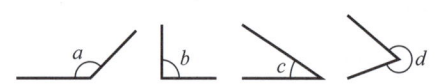

5 (a) Calculate angle e. _____ °

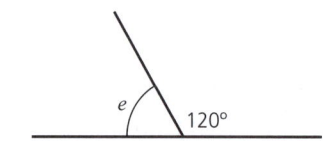

   (b) Calculate angle f. _____ °

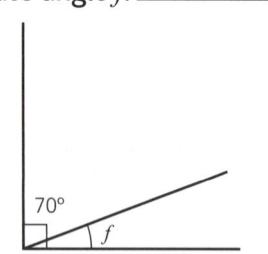

6 On the shape S below, side AB is parallel to side DC.

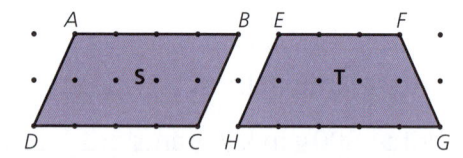

   (a) Name another pair of parallel sides of shape S. _____ and _____

   (b) Name a pair of parallel sides of shape T. _____ and _____

   (c) Name a pair of equal sides of shape T. _____ and _____

7 (a) Mark with stars [*] one pair of equal angles of shape S in question 6

   (b) Mark with stars one pair of equal angles of shape T in question 6

8 (a) On the co-ordinate grid, plot the points A (1, 3), B (3, 8) and C (5, 3).

   (b) Join the points in order to make shape ABC.

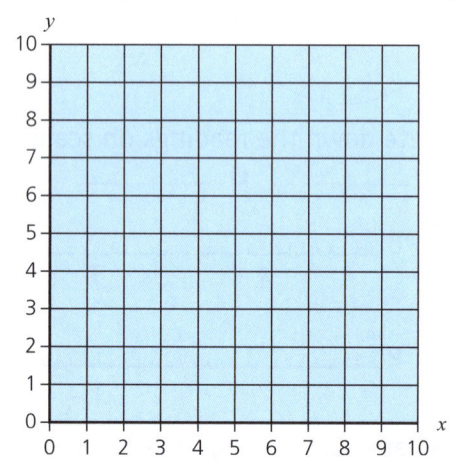

9 (a) Name the shape ABC that you have drawn in question 8 _____

   (b) Indicate with dashes [/] a pair of equal sides of shape ABC.

   (c) Indicate with arcs []] a pair of equal angles of shape ABC.

   (d) If shape ABC has symmetry, draw all the lines of symmetry.

10 On the grid, translate shape ABC five units to the right.

Shape, space and measures test 3

Final score
20 ___ %

1 (a) Measure the length of line **AB**, giving your answer to the nearest millimetre.

_____ mm

A _____ B

(b) Write 3 litres in millilitres. _____ ml

(c) Write 35 kg in grams. _____ g

2 (a) Which of the following is the best estimate of the mass of a drawing pin?

_____

1 mg   10 mg   100 mg   1 g   10 g

(b) Which of the following is the best estimate of the perimeter of this page?

_____

20 cm   40 cm   60 cm   1 m   2 m

3 (a) Write down the readings on scales **A** & **B**.

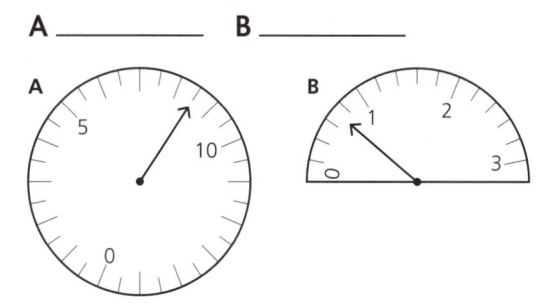

A _____   B _____

(b) Write down the readings on scales **C** & **D**

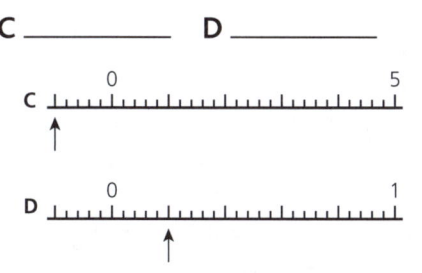

C _____   D _____

4 350 ml of water is poured into each cylinder.

Draw the water surface in each cylinder.

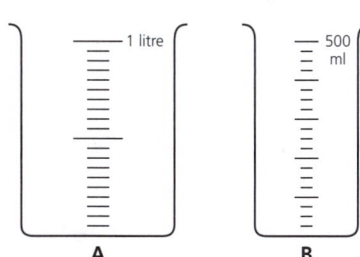

5 Look at the rectangle.

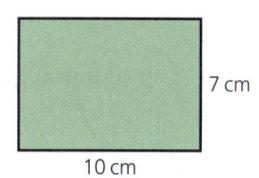

7 cm
10 cm

Calculate

(a) the perimeter _____ cm

(b) the area. _____ cm²

6 Look at the diagram.

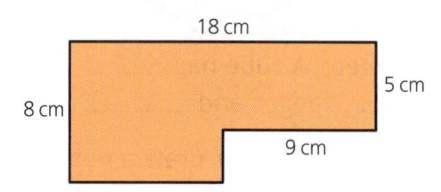

18 cm
5 cm
8 cm
9 cm

Calculate

(a) the perimeter _____ cm

(b) the area. _____ cm²

7 (a) Write 10.45 p.m. as a 24-hour time.

_____ : _____

(b) Write 12:28 as a 12-hour time. _____

(c) Sandy's birthday is 29 January and his sister Clare's birthday is 4 February. This year Clare's birthday is on a Saturday. On which day of the week will Sandy's birthday be this year? _____

8 Name the plane shapes below.

A _____   B _____

C _____   D _____

9 Draw all the lines of symmetry on the shapes **A, B, C** and **D** in question 8

10 Complete the shape below which is symmetrical about the line **m**.

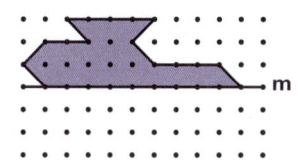

49

Shape, space and measures test 4

Final score
20 ___ %

1 (a) Name this polygon.
_____

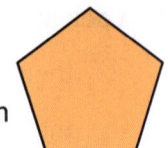

(b) What is the name of a polygon with 10 sides?
_____

2 The cube shown here is 2 cm long, 2 cm wide and 2 cm high.

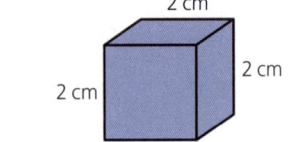

2 cm
2 cm
2 cm

(a) Complete: A cube has _____ faces, _____ edges and _____ vertices.

(b) On the grid below, draw a net for the cube. One face is drawn for you.

3 For the cube in question 2, what is

(a) the volume _____ cm³

(b) the area of card needed to make the cube? _____ cm²

4 Name the types of these angles.

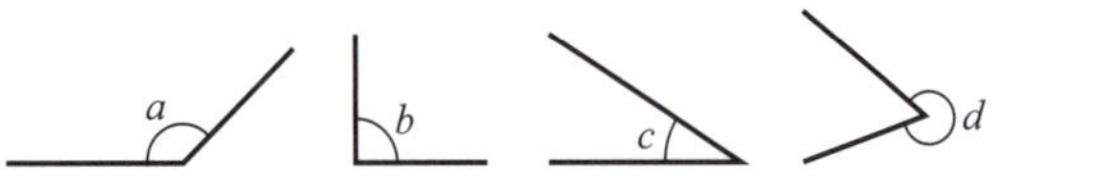

a _____ b _____ c _____ d _____

5 (a) Calculate angle e. _____ °

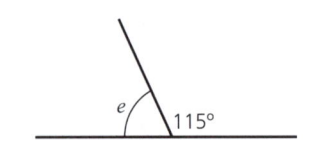

115°

(b) Calculate angle f. _____ °

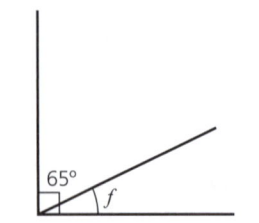

65°

6 On the shape S below, side AB is parallel to side DC.

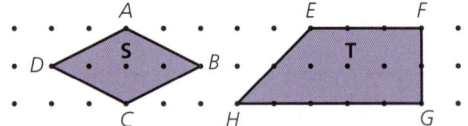

(a) Name another pair of parallel sides of shape S. _____ and _____

(b) Name a pair of parallel sides of shape T. _____ and _____

(c) Name a pair of equal sides of shape S. _____ and _____

7 (a) Mark with stars [*] one pair of equal angles of shape S in question 6

(b) Mark with stars one pair of equal angles of shape T in question 6

8 (a) On the co-ordinate grid, plot the points A (1, 6), B (3, 8), C (5, 6) and D (3, 2).

(b) Join the points in order to make shape ABCD.

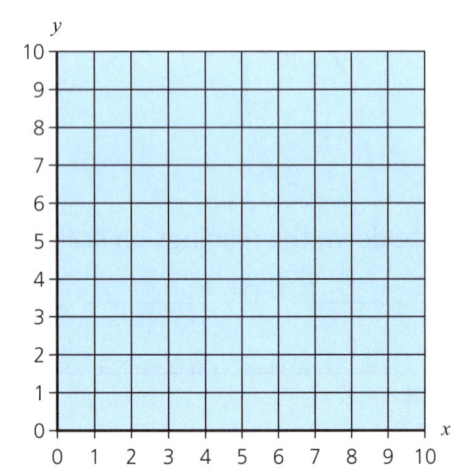

9 (a) Name the shape ABCD that you have drawn in question 8 _____

(b) Indicate with dashes [/] a pair of equal sides of shape ABCD.

(c) Indicate with arcs [)] a pair of equal angles of shape ABCD.

(d) If shape ABCD has symmetry, draw all the lines of symmetry.

10 On the grid, translate shape ABCD 4 units to the right.

Shape, space and measures test 5

Final score: 20 ___ %

1 (a) Measure the length of line *AB*, giving your answer to the nearest millimetre.
_____ mm

A _____ B

(b) Write 3.5 centimetres in millimetres.
_____ mm

(c) Write 3 g in milligrams. _____ mg

2 (a) Circle the area which is the best estimate of the area of a £5 note.
5 cm²   20 cm²   50 cm²   200 cm²   500 cm²

(b) Circle the capacity which is the best estimate of the capacity of a coffee mug.
10 ml   50 ml   100 ml   250 ml   500 ml

3 (a) Write down the readings on scales **A** & **B**.
A _____   B _____

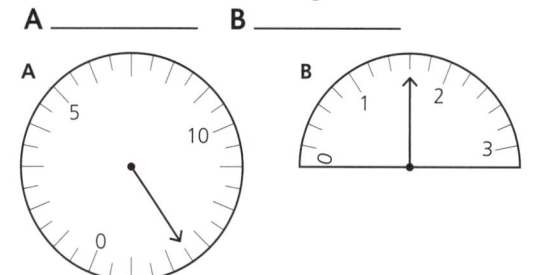

(b) Write down the readings on scales **C** & **D**.
C _____   D _____

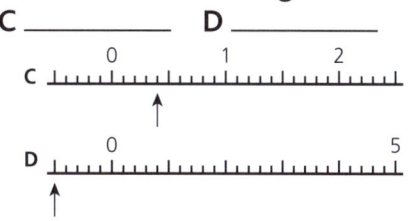

4 All the water in cylinder **A** is poured into cylinder **B**.
Draw the new water surface in cylinder **B**.

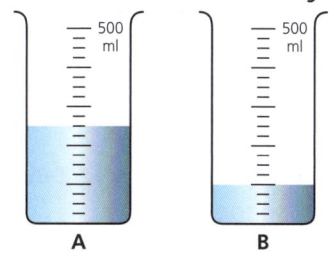

5 The diagram shows a rectangle.

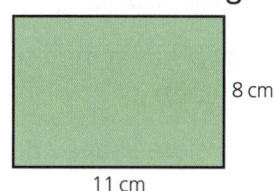

(8 cm, 11 cm)

Calculate
(a) the perimeter _____ cm
(b) the area. _____ cm²

6 Look at the diagram.

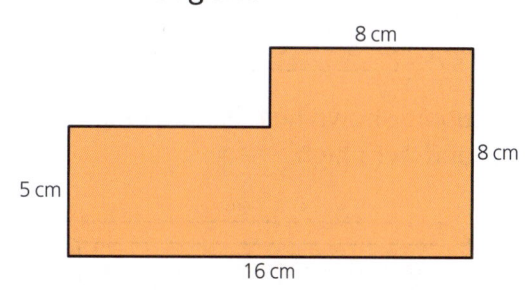

(8 cm, 8 cm, 5 cm, 16 cm)

Calculate
(a) the perimeter _____ cm
(b) the area. _____ cm²

7 (a) Write 11.05 p.m. as a 24-hour time.
____ : ____

(b) Write 11:05 as a 12-hour time. _____

(c) Fay's birthday is 25 June and her sister Kelly's birthday is 10 days later. This year Fay's birthday is on a Wednesday. On which day of the week will Kelly's birthday be this year? _____

8 Name the plane shapes below.
A _____   B _____
C _____   D _____

9 Draw all the lines of symmetry on the shapes **A, B, C** and **D** in question 8

10 Complete the shape below which is symmetrical about the line **m**.

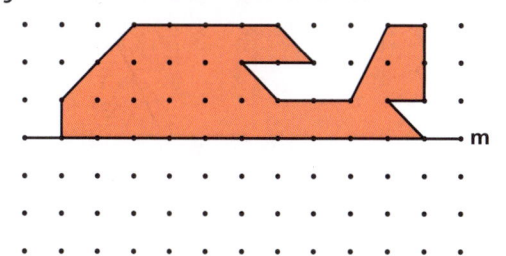

51

Shape, space and measures test 6

Final score
20 ___ %

1 (a) Name this polygon. _____

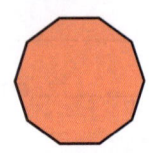

   (b) What is the name of a polygon with 8 sides? _____

2 The cuboid shown here is 7 cm long, 1 cm wide and 1 cm high.

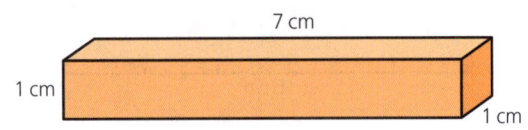

   (a) Complete: A cuboid has _____ faces, _____ edges and _____ vertices.

   (b) On the square dotted grid draw a net for the cuboid. One face is drawn for you.

3 For the cuboid in question 2, what is

   (a) the volume _____ cm³

   (b) the area of card needed to make the cuboid?

   _____ cm²

4 Name the types of these angles. a _____
   b _____ c _____ d _____

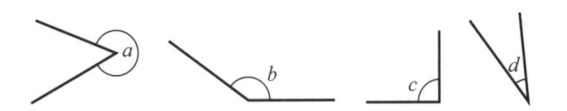

5 Calculate angles e and f. e _____° f _____°

   51°   63°

6 On the shape S below, side AD is parallel to side BC.

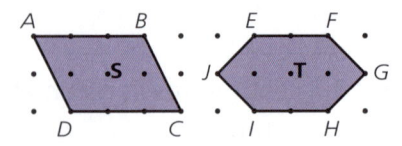

   (a) Name another pair of parallel sides of shape S. _____ and _____

   (b) Name a pair of parallel sides of shape T. _____ and _____

   (c) Name a pair of equal sides of shape T. _____ and _____

7 (a) Mark with stars [*] one pair of equal angles of shape S in question 6

   (b) Mark with stars one pair of equal angles of shape T in question 6

8 (a) On the co-ordinate grid, plot the points A (2, 4), B (4, 6), C (6, 4) and D (4, 2).

   (b) Join the points in order to make shape ABCD.

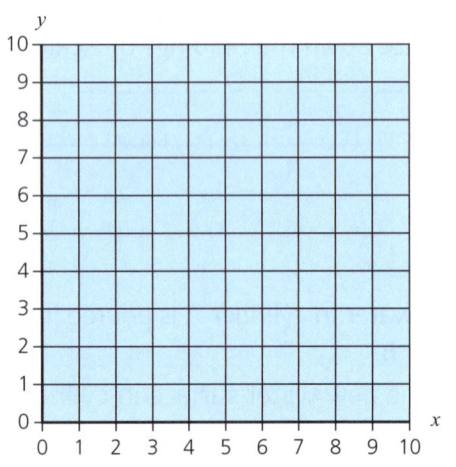

9 (a) Name the shape ABCD that you have drawn in question 8 _____

   (b) Indicate with dashes [/] a pair of equal sides of shape ABCD.

   (c) Indicate with arcs [)] a pair of equal angles of shape ABCD.

   (d) If shape ABCD has symmetry, draw all the lines of symmetry.

10 On the grid, translate shape ABCD 3 units to the right and 3 units up.

Shape, space and measures test 7

Final score
20 ___ %

1 (a) Measure the length of line AB, giving your answer to the nearest millimetre.
   _____ mm
   A _____ B

   (b) Write 3 millimetres in centimetres.
   _____ cm

   (c) Write 450 g in kilograms. _____ kg

2 (a) Circle the measurement which is the best estimate of the distance round the edge of a normal dinner plate.

   3 cm   10 cm   30 cm   100 cm   300 cm

   (b) Circle the measurement which is the best estimate of the area of your hand.

   10 cm²   20 cm²   40 cm²   80 cm²   160 cm²

3 (a) Write down the readings on scales A & B.
   A _____   B _____

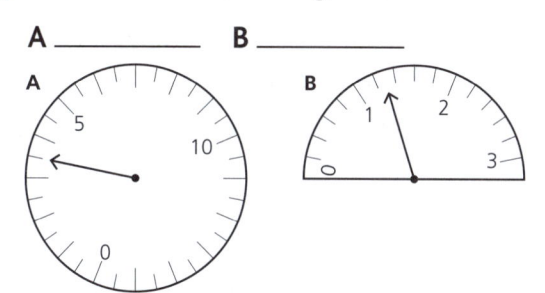

   (b) Write down the readings on scales C & D.
   C _____   D _____

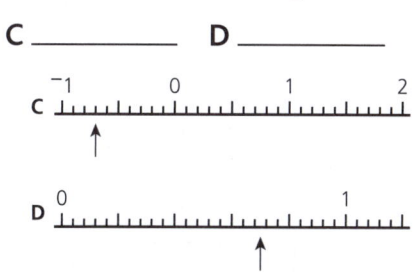

4 All the water in cylinder A is poured into cylinder B.

   Draw the water surface in cylinder B.

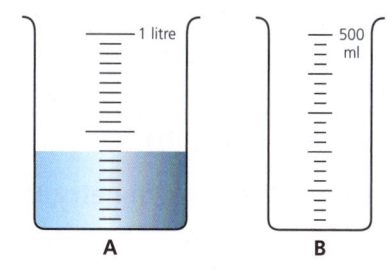

5 The diagram shows a rectangle.

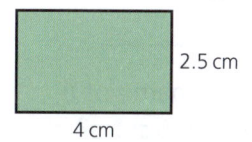

   2.5 cm
   4 cm

   Calculate

   (a) the perimeter _____ cm

   (b) the area. _____ cm²

6 Look at the diagram.

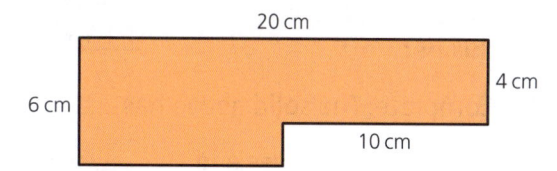

   20 cm
   4 cm
   6 cm
   10 cm

   Calculate

   (a) the perimeter _____ cm

   (b) the area. _____ cm²

7 (a) Write 5.45 p.m. as a 24-hour time.
   _____ : _____

   (b) Write 21:35 as a 12-hour time. _____

   (c) Kirsten's birthday is 28 January and her mother's birthday is 7 days later. This year her mother's birthday is on a Sunday. On which day of the week will Kirsten's birthday be this year? _____

8 Name the plane shapes below.
   A _____   B _____
   C _____   D _____

9 Draw all the lines of symmetry on the shapes A, B, C and D in question 8

10 Complete the shape below which is symmetrical about the line m.

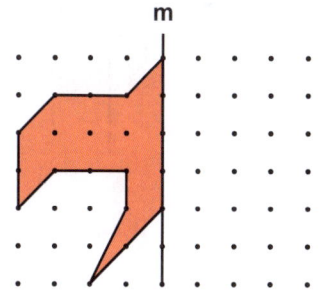

Shape, space and measures test 8

Final score
20 ___ %

1 (a) Name this polygon.
   _____

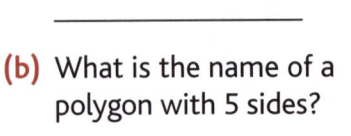

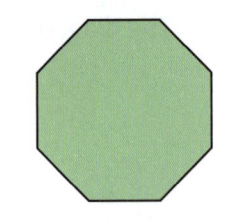

  (b) What is the name of a polygon with 5 sides?
   _____

2 The net of a solid shape is drawn on this grid.

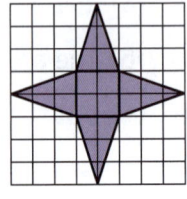

  (a) Name the shape that can be made from this net. _____

  (b) Complete: The solid above has ____ faces, ____ edges and ____ vertices.

3 A cuboid measures 5 cm by 4 cm by 3 cm. Calculate

  (a) the volume _____ cm³

  (b) the area of card needed to make the cuboid. _____ cm²

4 Name the types of these angles. a _____
  b _____ c _____ d _____

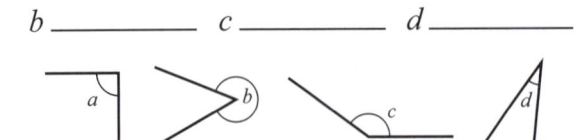

5 (a) Calculate angle e. _____ °

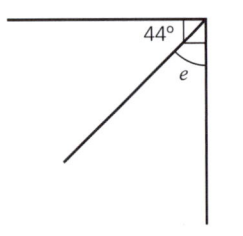

  (b) Calculate angle f. _____ °

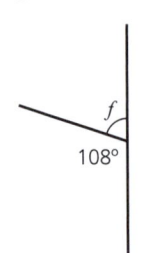

6 On the shape S below, side AB is parallel to side DC.

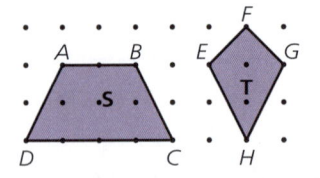

  (a) Name a pair of equal sides of shape S.
   _____ and _____

  (b) Name two pairs of equal sides of shape T.
   _____ and _____
   _____ and _____

7 (a) Mark with stars [*] one pair of equal angles of shape S in question 6

  (b) Mark with stars one pair of equal angles of shape T in question 6

8 (a) On the grid, plot the points A (1, 6), B (6, 9), C (9, 9) and D (4, 6).

  (b) Join the points in order to make shape ABCD.

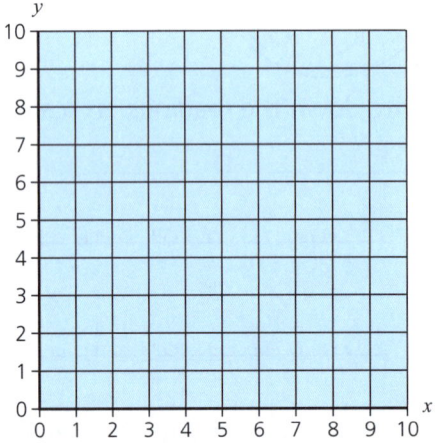

9 (a) Name the shape ABCD that you have drawn in question 8 _____

  (b) Indicate with dashes [/] a pair of equal sides of shape ABCD.

  (c) Indicate with arcs [)] a pair of equal angles of shape ABCD.

  (d) If shape ABCD has symmetry, draw all the lines of symmetry.

10 On the grid, translate shape ABCD five units down.

Shape, space and measures test 9

Final score
20 ___ %

1 Measure the perimeter of the shape below, giving the answer to the nearest millimetre.
   _____ mm

2 (a) Circle the area which is the best estimate of the mass of this book.

   500 mg   5 g   50 g   500 g   5 kg   50 kg

   (b) Circle the best estimate of the length of an adult mouse (including its tail).

   10 mm   5 cm   10 cm   50 cm   1 m

3 (a) Write down the readings on scales **A** & **B**.
   A _____   B _____

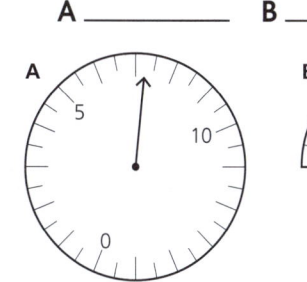

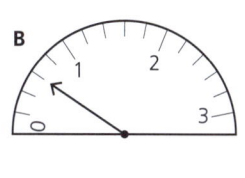

   (b) Write down the readings on scales **C** & **D**.
   C _____   D _____

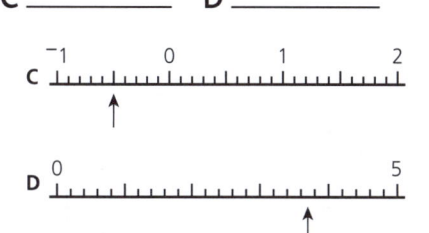

4 (a) What is the volume of water in the cylinder? _____ ml

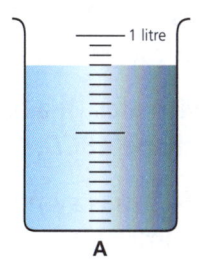

   (b) How many 220 ml glasses could be filled from the cylinder and how much water will be left over? _____ glasses, _____ ml left

5 A rectangle is twice as long as it is wide. The perimeter is 18 cm.

   What is the area? _____ cm²

6 This shape has been made from a 4 cm square by cutting a square of side 1 cm from each corner.

   Calculate

   (a) the perimeter _____ cm

   (b) the area. _____ cm²

7 (a) Write 7.50 p.m. as a 24-hour time.
   _____ : _____

   (b) Write 14:55 as a 12-hour time. _____

   (c) Sam's birthday is 28 September and his sister Alice's birthday is 7 days later. This year Alice's birthday is on a Sunday. On which day of the week will Sam's birthday be this year? _____

8 Name the plane shapes below.
   A _____   B _____
   C _____   D _____

9 Draw all the lines of symmetry on the shapes **A**, **B**, **C** and **D** in question 8

10 Complete the shape below which is symmetrical about the line **m**.

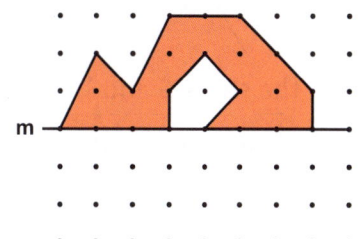

Shape, space and measures test 10

Final score: 20 ___ %

1 (a) Name this polygon. _____

(b) What is the name of a polygon with 10 sides? _____

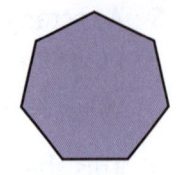

2 The diagram shows a hexagonal prism made from transparent plastic. The top and bottom faces are regular hexagons.

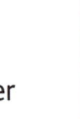

(a) What shape are the other 6 faces? _____

(b) Complete: A hexagonal prism has ___ faces, ___ edges and ___ vertices.

3 A cuboid is 4 cm long, 1 cm wide and 2.5 cm high. Calculate

(a) the volume _____ cm³

(b) the area of card needed to make the cuboid. _____ cm²

4 Name the types of these angles. a _____ b _____ c _____ d _____

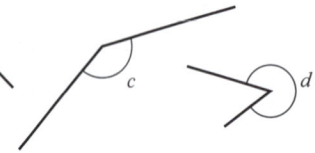

5 (a) Calculate angle e. _____ °

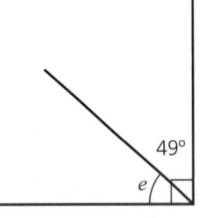

(b) Calculate angle f. _____ °

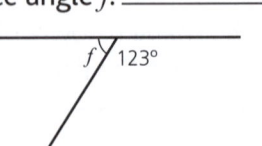

6 On each of the shape S and T

(a) mark with stars a pair of equal angles

(b) mark with dashes (/) a pair of equal sides.

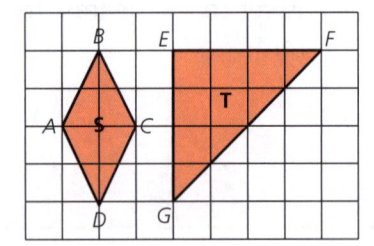

7 On the square dotted grid, draw

(a) a quadrilateral with just one pair of parallel sides

(b) a quadrilateral with just two equal angles and two pairs of equal sides.

8 (a) On the co-ordinate grid, plot points A (2, 6), B (4, 9), C (6, 6) and D (4, 3).

(b) Join the points in order to make shape ABCD.

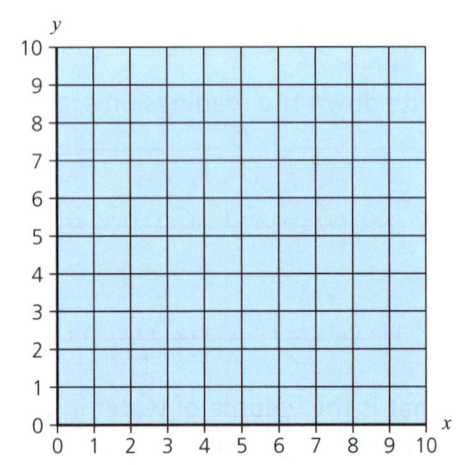

9 (a) Name the shape ABCD that you have drawn in question 8 _____

(b) Indicate with arcs [)] a pair of equal angles of shape ABCD.

(c) If shape ABCD has symmetry, draw all the lines of symmetry.

10 On the grid, translate shape ABCD 3 units to the right and 3 units down.

56

Handling data test 1

Final score
20 ___ %

Questions 1 to 4 refer to the table below of data for five friends, all born in the same year.

Name	Birthday	Number of pets	Favourite sport
Anne	22 November	2	hockey
Daniel	06 December	0	football
Sofia	23 January	4	hockey
Dylan	29 November	3	football
Ellie	12 July	2	hockey

1 (a) What is the total number of pets that the 5 friends have? _____

   (b) What number of pets is the mode? _____

2 (a) Who is the oldest? _____

   (b) When the children are arranged in order of age, who will be in the middle? _____

3 Who is a week older than Dylan? _____

4 (a) What fraction of the friends has no pets? _____

   (b) What percentage of the friends chose hockey as their favourite sport? _____ %

Questions 5 to 7 refer to the pictogram below which records the sales of ice-creams.

One symbol represents one ice-cream.

Vanilla	🍦🍦🍦🍦🍦
Raspberry	🍦🍦🍦🍦
Mint	🍦
Chocolate	🍦🍦🍦🍦🍦🍦🍦
Strawberry	🍦🍦

5 (a) Which was the most popular flavour of ice-cream? _____

   (b) How many more vanilla ice-creams were sold than mint ice-creams? _____

6 What was the total number of ice-creams sold? _____

7 (a) What fraction (in its simplest form) of the ice-creams sold was raspberry flavour? _____

   (b) What percentage of the ice-creams sold was chocolate flavour? _____ %

Questions 8 to 10 concern rolling a die.

The tally below records the scores when the die was rolled 45 times.

Score	Tally of marks	Frequency
1									
2							
3									
4										
5						
6										

The die is rolled another five times with the following scores:

1 5 2 1 4

8 Add these five scores to the tally above.

9 Complete the frequency table above.

10 Complete the diagram below.

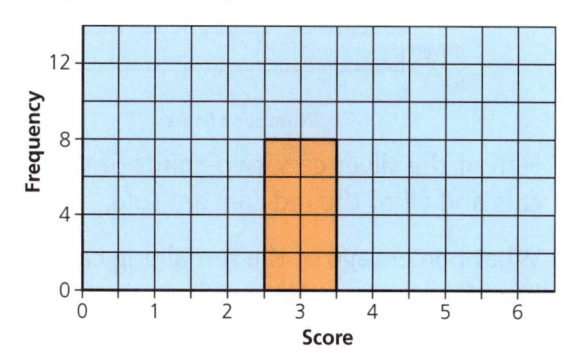

Handling data test 2

Final score
20 ___ %

Questions 1 and 2 refer to the letters below.

A B C D E F G H I J

1. Write the letters in the correct regions of the Venn diagram below.

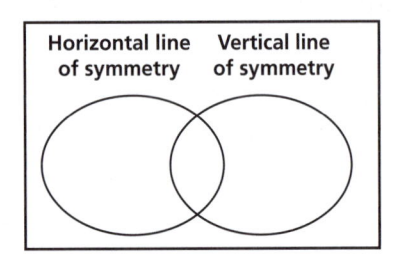

   Horizontal line of symmetry | Vertical line of symmetry

2. Write the letters in the Carroll diagram.

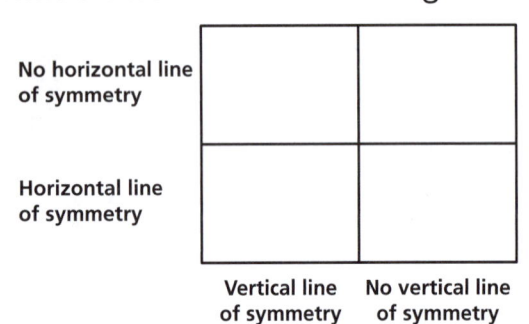

	Vertical line of symmetry	No vertical line of symmetry
No horizontal line of symmetry		
Horizontal line of symmetry		

Questions 3 to 5 refer to the diagram below showing the colours of 20 cars in a showroom.

3. (a) What is the most common colour? _____

   (b) What fraction of the cars is black? _____

4. Represent this information as a bar chart.

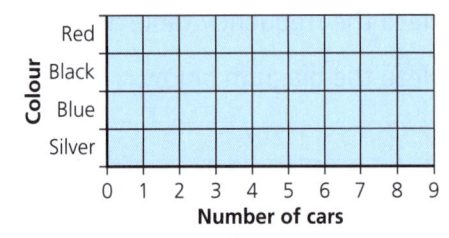

5. Half of the silver cars, two-thirds of the blue cars and all of the red cars are sold.

   What percentage of the remaining cars is black? _____ %

Questions 6 and 7 refer to the line graph below which a shopkeeper uses when selling stickers.

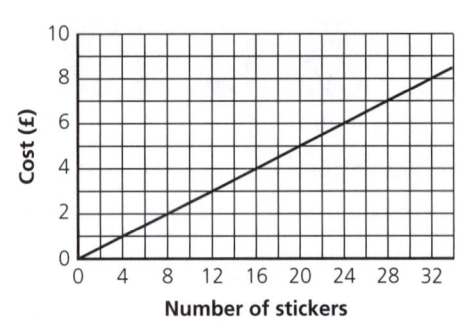

6. (a) What is the cost of 4 stickers? £ _____

   (b) How many stickers could be bought for £5? _____

7. (a) What is the cost of 22 stickers? £ _____

   There are 48 stickers in the complete set.

   (b) What would be the cost of one complete set? £ _____

8. Tick which is more likely:

   - scoring 3 when an ordinary die is rolled
   - getting 'heads' when a coin is tossed.

9. On the scale below, which letter best represents the likelihood of

   (a) snow falling somewhere in Britain in January _____

   (b) you growing to be 6 m tall? _____

   A   B   C   D   E   F   G
   Impossible   Even chance   Certain

10. On the scale above, which letter best represents the likelihood of getting

    (a) an even number when a die is rolled _____

    (b) 'tails' when a coin is tossed? _____

58

Handling data test 3

Final score
20 ____ %

Questions 1 to 4 refer to the table below of data for five friends, all born in the same year.

Name	Birthday	Height (cm)	Hair colour
Freya	31 July	140	brown
Natalie	06 January	135	black
Leon	29 March	143	brown
Isla	08 December	139	blonde
Jane	31 August	141	blonde

1 (a) Who is the oldest? _____

   (b) How many of the friends are taller than Freya? _____

2 (a) Whose birthday is closest to Christmas Day? _____

   (b) When the friends are arranged in order of height, who will be in the middle? _____

3 How many days older than Jane is Freya? _____

4 (a) What fraction of the friends has blonde hair? _____

   (b) What percentage of the friends has brown hair? _____%

Questions 5 to 7 refer to the pictogram below which shows the number of bears sold by a shop.

One symbol 🧸 represents one bear.

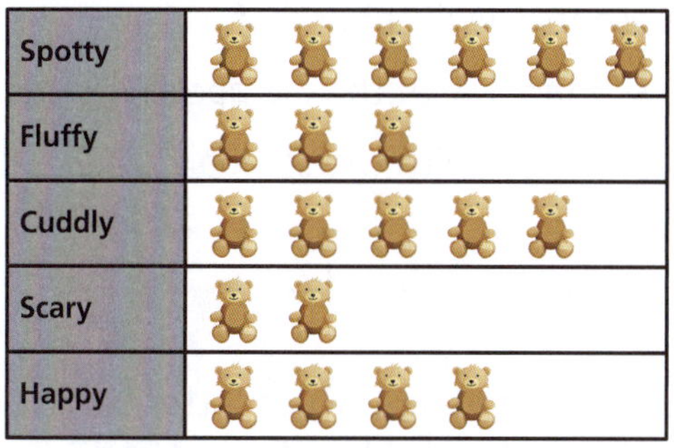

Spotty	🧸 🧸 🧸 🧸 🧸 🧸
Fluffy	🧸 🧸 🧸
Cuddly	🧸 🧸 🧸 🧸 🧸
Scary	🧸 🧸
Happy	🧸 🧸 🧸 🧸

5 (a) Which was the most popular bear? _____

   (b) How many more Cuddly bears were sold than Scary bears? _____

6 What was the total number of bears sold? _____

7 (a) What fraction (in its simplest form) of the bears sold was Cuddly bears? _____

   (b) What percentage of the bears sold was Happy bears? _____%

Questions 8 to 10 concern spinning a spinner.

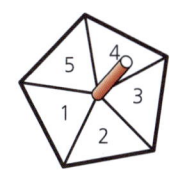

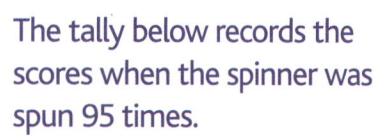

The tally below records the scores when the spinner was spun 95 times.

Score	Tally of marks	Frequency
1	𝍢 𝍢 𝍢 𝍢 II	
2	𝍢 𝍢 𝍢 II	
3	𝍢 𝍢 𝍢 III	18
4	𝍢 𝍢 𝍢 IIII	
5	𝍢 𝍢 𝍢 IIII	

The spinner is spun another five times with the following scores: 4 2 5 1 2

8 Add these five scores to the tally above.

9 Complete the frequency table above.

10 Complete the chart below.

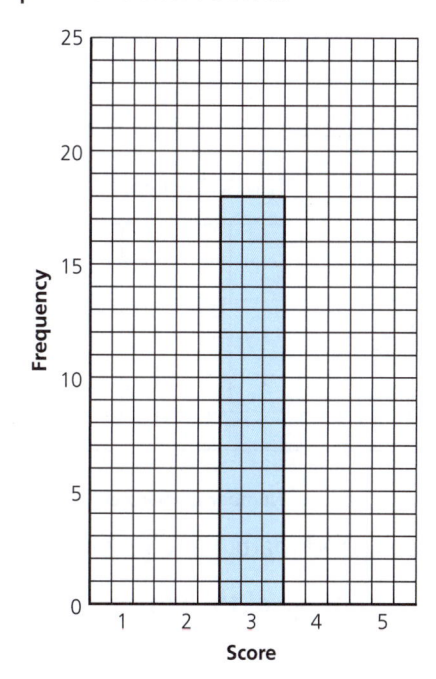

59

Handling data test 4

Final score
20 ___ %

Questions 1 and 2 refer to the numbers below.

1 2 3 4 5 6 7 8 9

1. Complete the Venn diagram by writing the numbers in the correct regions. The first three numbers have been put in for you.

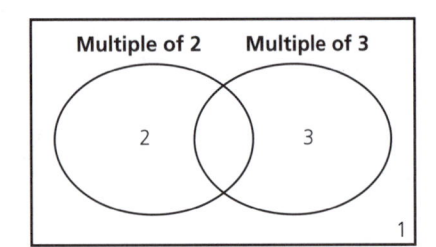

   Multiple of 2 — Multiple of 3
   2, 3, 1

2. Write the numbers in the Carroll diagram.

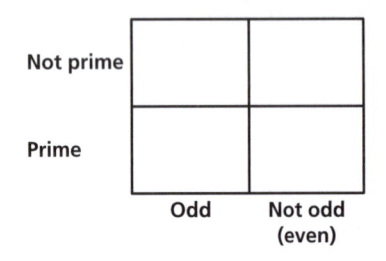

	Odd	Not odd (even)
Not prime		
Prime		

Questions 3 to 5 refer to the diagram below showing the colours of Tim's 12 marbles.

3. What fraction of the marbles (in its simplest form) is

   (a) blue _____

   (b) red? _____

4. Represent this information as a bar chart.

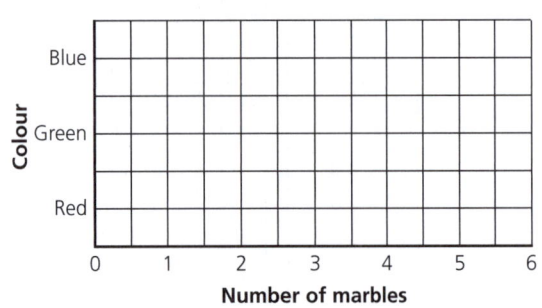

5. Tim loses a red marble and green marble.

   What percentage of the remaining marbles is green. _____ %

Questions 6 and 7 refer to the line graph which Jane uses to calculate the distance she can walk (at a steady pace) in different lengths of time.

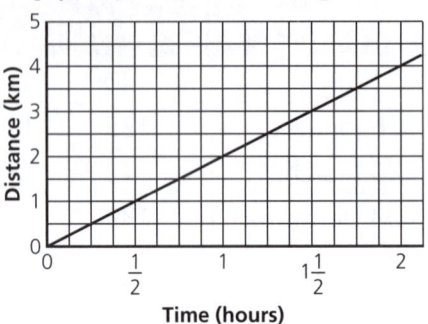

6. How far can Jane walk in

   (a) 1 hour _____ km

   (b) 45 minutes? _____ km

7. How long would it take Jane to walk

   (a) 4 kilometres _____ hours

   (b) $3\frac{1}{2}$ kilometres? _____ hour(s) _____ minutes

Questions 8 to 10 concern the spinner and die shown here.

8. Tick which is more likely:
   - scoring 1 when the spinner is spun
   - getting 1 when the die is rolled.

9. On the scale below, which letter best represents the likelihood of

   (a) scoring 5 with the die above _____

   (b) scoring 5 with the spinner? _____

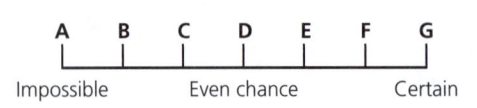

   A   B   C   D   E   F   G
   Impossible   Even chance   Certain

10. On the scale above, which letter best represents the likelihood of getting

    (a) an odd number when the die is rolled _____

    (b) a number less than 5 when the spinner is spun? _____

60

Handling data test 5

Final score
20 ___ %

Questions 1 to 4 refer to the table below of data for five friends, all born in the same year.

Name	Age (years : months)	Mass (kg)	Eye colour
Kiera	9 : 10	39	Blue
Liam	9 : 8	35	Brown
Maria	9 : 11	40	Brown
Nicole	9 : 0	38	Blue
Oliver	9 : 7	36	Blue

1 (a) Who was born first? _____

   (b) How many of the friends are older than Oliver? _____

2 (a) Who has the smallest mass? _____

   (b) When the friends are arranged in order of increasing mass, who will be in the middle? _____

3 Who has brown eyes and is older than Kiera? _____

4 (a) What fraction of the friends has blue eyes? _____

   (b) What percentage of the friends has brown eyes? _____%

Questions 5 to 7 refer to the pictogram below which records the money raised by 5 friends.

One symbol ◯ represents two pounds.

Peter	◯◯◯◯◖
Riley	◯◯◯◯◯◯◯
Rebecca	◯◯◯◯◯◯◖
Sienna	◯◯
Tammy	◯◯◯◯

5 (a) How much money did Peter raise? £_____

   (b) How much more did Riley raise than Rebecca? £_____

6 What was the total amount of money raised? £_____

7 (a) What percentage of the total amount was raised by Rebecca? _____%

   (b) Rebecca shared her money equally between two charities. How much did she give to each charity? £_____

Questions 8 to 10 concern spinning a spinner.

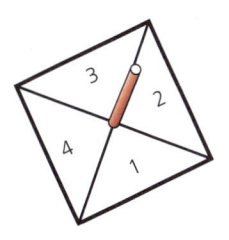

The tally below records the scores when the spinner was spun.

Score	Tally of marks	Frequency
1	llll llll ll	12
2	llll llll l	
3	llll llll llll	
4	llll llll ll	

8 (a) Complete the table of frequencies.

   (b) How many times was the spinner spun? _____

9 What percentage of the spins resulted in a score of 4? _____%

10 Complete the frequency diagram below.

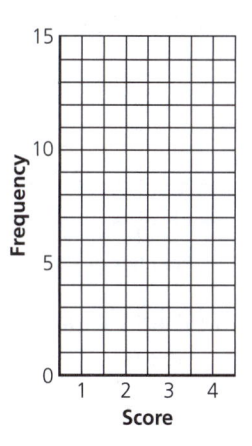

61

Handling data test 6

Final score
20 ___ %

Questions 1 and 2 refer to the numbers below.

10 11 12 13 14 15 16 17 18 19

1 Write the numbers in the correct regions of the Venn diagram below.

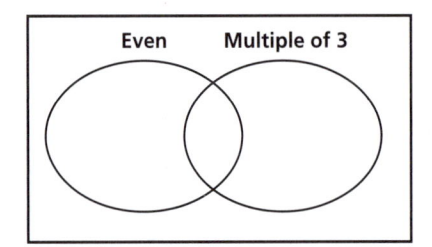

[Venn diagram: Even | Multiple of 3]

2 Write the numbers in the Carroll diagram.

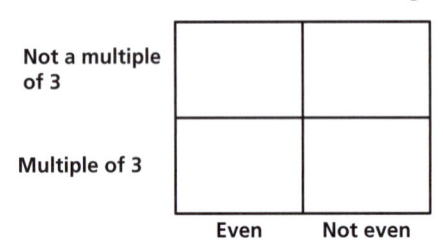

[Carroll diagram: rows = Not a multiple of 3, Multiple of 3; columns = Even, Not even]

Questions 3 to 5 refer to the diagram below showing the colours of chocolates in a box.

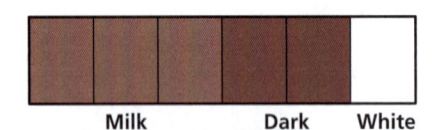

[Diagram showing Milk, Dark, White chocolates]

In the box there are 4 dark chocolates.

3 (a) How many white chocolates are there? _____

   (b) What is the total number of chocolates? _____

4 Represent this information as a bar chart.

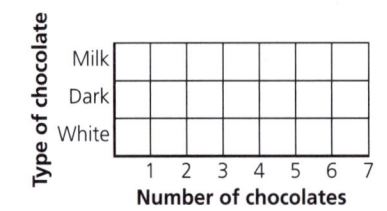

[Blank bar chart: Type of chocolate (Milk, Dark, White) vs Number of chocolates 1–7]

5 What fraction (in its simplest form) of the chocolates in a box is

   (a) milk _____

   (b) white? _____

Questions 6 and 7 refer to the line graph below which can be used convert between inches and centimetres.

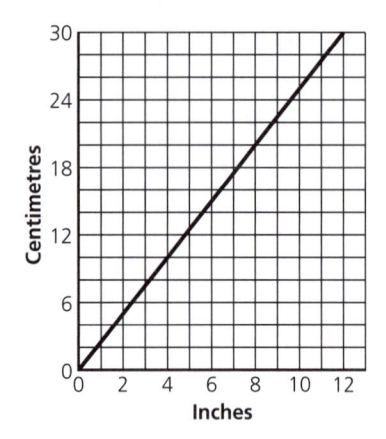

[Line graph: Centimetres (0–30) vs Inches (0–12)]

6 (a) Change 4 inches to centimetres. _____ cm

   (b) Change 20 cm to inches. _____ inches

7 (a) How many centimetres are there to 1 foot (12 inches)? _____ cm

   (b) Mr Smith is 6 feet tall. What is his height in centimetres? _____ cm

8 Tick which is more likely:
   • scoring a number less than 4 when an ordinary die is rolled
   • scoring a number greater than 4 when an ordinary die is rolled.

9 On the scale below, which letter best represents the likelihood of

   (a) you blinking your eyes in the next ten minutes _____

   (b) you gaining 10 kg in mass during the next week? _____

   A   B   C   D   E   F   G
   Impossible   Even chance   Certain

10 On the scale above, which letter best represents the likelihood, when an ordinary die is rolled, of getting

   (a) a prime number _____

   (b) a square number? _____

Handling data test 7

Final score
20 ___ %

Questions 1 to 4 refer to the table below of data for six friends.

Name	Number of brothers	Number of sisters	Number of pets
Una	1	0	3
Violet	2	2	2
Olivia	1	2	4
Jayden	0	1	2
Eric	3	0	6
Zac	1	2	1

1 (a) Who has most pets? _____

  (b) Who has most brothers and sisters? _____

2 (a) How many pets do the six friends have in total? _____

  (b) What is the most common number of pets? _____

3 How many of the friends have more brothers than sisters? _____

4 (a) What fraction of the friends has no sisters? _____

  (b) What percentage of the friends has 3 or more pets? _____%

Questions 5 to 7 refer to the pictogram below showing the number of drinks sold by a café.

One symbol 🍵 represents ten drinks.

Tea	🍵 🍵 🍵 🍵 🍵 🥃
Lemon tea	🍵 🍵 🥃
Cappuccino	🍵 🍵 🍵 🥃
Latte	🍵 🍵 🍵 🍵 🍵
Espresso	🍵 🍵 🍵 🥃

5 (a) How many cups of tea were sold? _____

  (b) What was the total number of drinks sold? _____

6 If all of the drinks are priced at £1.20 each, how much money did the café take?
   £ _____

7 (a) What fraction (in its simplest form) of the drinks sold was lemon tea? _____

  (b) What percentage of the drinks sold was latte? _____%

Questions 8 to 10 concern tossing a coin. The results are written on a piece of paper as shown below.

H	T	H	T	H	H	H	T	H	H
T	T	H	H	T	T	H	H	H	T

8 (a) How many times has the coin been tossed? _____

  (b) Complete the tally column below.

Score	Tally of marks	Frequency
'heads' (H)		
'tails' (T)		

9 (a) Complete the frequency column above.

  (b) What percentage of the tosses resulted in 'heads'? _____%

10 Represent the results on the diagram below.

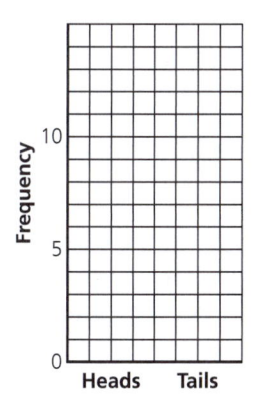

Frequency
10
5
0
Heads Tails

Handling data test 8

Final score
20 ___ %

Questions 1 and 2 refer to the set of twelve letter cards below.

[A] [C] [D] [E] [F] [H] [I] [L] [M] [N] [O] [R]

1 Marie and Nicole take it in turn to make their names.

[M][A][R][I][E] and [N][I][C][O][L][E]

Write the letters on the set of 12 cards in the correct regions of the Venn diagram.

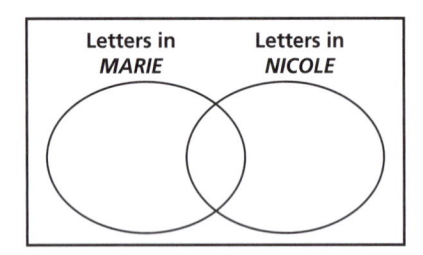

Letters in MARIE | Letters in NICOLE

2 Write the letters on the set of 12 cards in the Carroll diagram.

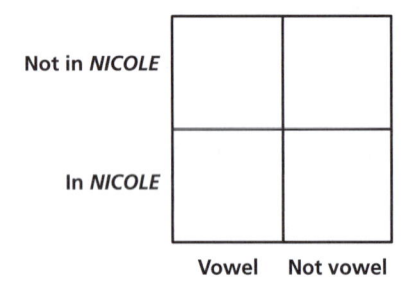

	Vowel	Not vowel
Not in NICOLE		
In NICOLE		

Questions 3 to 5 refer to the diagram below showing information about pupils in Year 5

Boys | Girls

There are 10 boys in Year 5

3 What percentage of the pupils is girls?
   _____%

4 How many pupils are there in Year 5?
   _____

5 Half of the boys and two-thirds of the girls in Year 5 sing in the choir. What percentage of the pupils in Year 5 sings in the choir?
   _____%

Questions 6 and 7 refer to the line graph below which shows the length of Fraser's pencil measured at the start of each day.

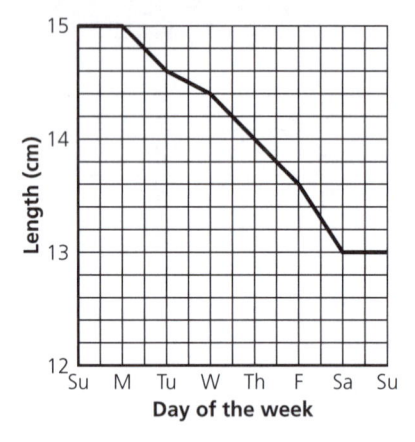

6 How long was the pencil at the start of Friday?

   _____ cm

7 On which day did Fraser do most writing?

   _____

Questions 8 to 10 concern the die and spinner shown here.

8 Tick which is more likely:
   • getting blue when the spinner is spun
   • getting an odd number when the die is rolled.

9 On the scale below, which letter best represents the likelihood of getting a number greater than 4 when the die is rolled?

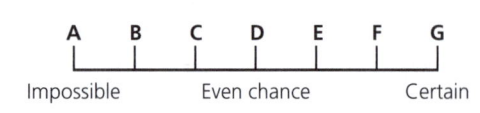

   _____

   A   B   C   D   E   F   G
   Impossible   Even chance   Certain

10 When the spinner is spun, the likelihood of getting blue lies between which two letters on the scale above?

   ____ and ____

Handling data test 9

Final score
20 ___ %

Questions 1 to 4 refer to the table below of data for seven friends.

Name	Age (y : m)	Mass (kg)	Height (cm)	Favourite colour
Amy	9 : 8	39	121	Blue
Ben	9 : 9	35	118	Yellow
Haley	9 : 7	40	120	Pink
Dale	10 : 1	38	115	Blue
Emma	9 : 11	36	119	Pink
Freddie	10 : 0	37	114	Red
Mia	9 : 10	39	123	Pink

1 (a) Who is the youngest? _____

   (b) How many months older than Amy is Dale? _____

2 (a) Who has the smallest mass? _____

   (b) When the friends are arranged in order of increasing height, who will be in the middle? _____

3 Who is heavier than Freddie, taller than Amy and has favourite colour pink? _____

4 (a) What fraction of the friends is aged 10? _____

   (b) What percentage of the 9 year olds has favourite colour pink? _____%

Questions 5 to 7 refer to the pictogram below showing the results of a ladybird-finding study.

One symbol 🐞 represents two ladybirds.

Type A	🐞 🐞 🐞 🐞 🐞
Type B	🐞 🐞 🐞
Type C	🐞 🐞 🐞
Type D	🐞 🐞 🐞 🐞 🐞 🐞

5 (a) How many beetles of type B were found? _____

   (b) How many more beetles of type A were found than beetles of type B? _____

6 How many beetles were found in total? _____

7 What percentage of the beetles found was of type C? _____%

Questions 8 to 10 concern a traffic survey.

The tally records the numbers of vehicles passing the gate between 12:00 and 12:15

Vehicle	Tally	Frequency
Car	𝍬𝍬𝍬𝍬𝍬III	28
Van	𝍬𝍬𝍬𝍬IIII	
HGV	𝍬𝍬𝍬𝍬I	
Other	𝍬𝍬𝍬II	

8 (a) Complete the frequency table.

   (b) What was the total number of vehicles that passed the gate? _____

9 How many vehicles might you expect to pass the gate in an hour at lunchtime? _____

10 The frequency diagram below shows the numbers of vehicles passing the same gate in **10 minutes** in the middle of the afternoon.

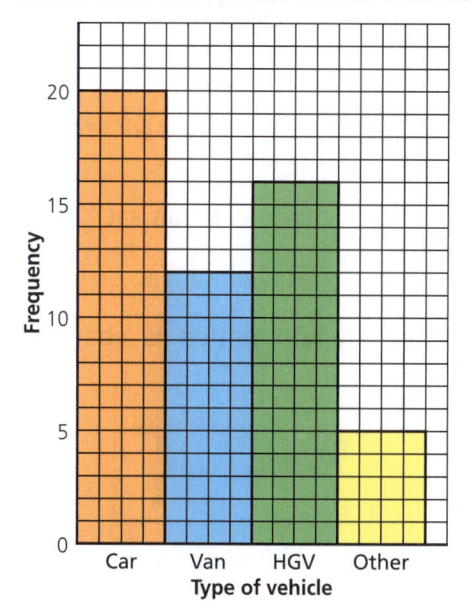

   (a) How many vehicles in total passed the gate during this 10 minute period? _____

   (b) Estimate, to the nearest 50, how many vehicles might pass the gate in an hour during the middle of the afternoon. _____

65

Handling data test 10

Final score
20 ___ %

Questions 1 and 2 refer to the numbers below.

20 21 22 23 24 25 26 27 28 29

1. Write the numbers in the correct regions of the Venn diagram below.

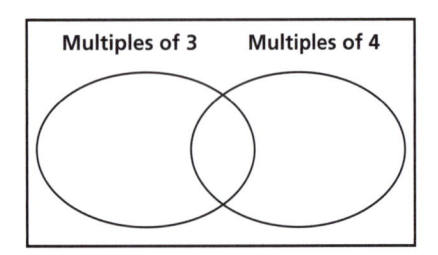

   Multiples of 3   Multiples of 4

2. Write the numbers in the Carroll diagram.

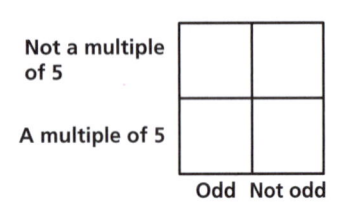

	Odd	Not odd
Not a multiple of 5		
A multiple of 5		

Questions 3 to 5 refer to the diagram below showing the 8 slices of a pizza.

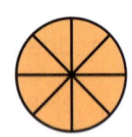

3. Lindsay eats one slice of the pizza and Jamie eats $\frac{3}{7}$ of what is left.

   How many slices does Jamie eat? _____

Jamie's dog, Fido, eats half of the remaining pizza.

4. Represent the information as a bar chart.

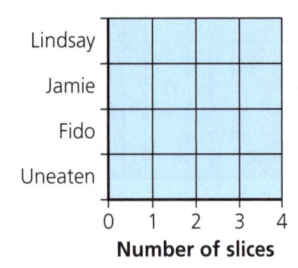

   Lindsay
   Jamie
   Fido
   Uneaten
   0 1 2 3 4
   Number of slices

5. What fraction of the pizza remains uneaten?

   _____

Questions 6 and 7 refer to the line graph below showing the amount of money in Tom's account.

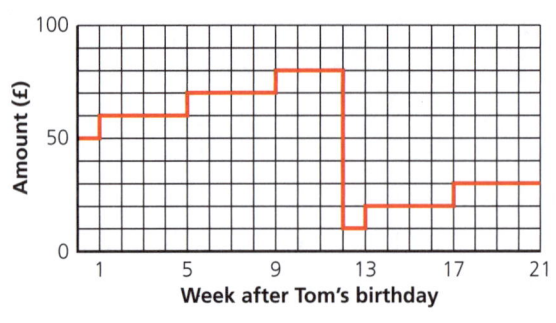

Amount (£)
100
50
0
   1   5   9   13   17   21
Week after Tom's birthday

Tom starts to keep a record on his 10th birthday. Starting one week after his birthday, he receives a fixed amount of pocket money, paid into his account every 4 weeks.

6. Which week did Tom take money out of his account and how much money did he take out?

   Week _____ £ _____

7. If Tom doesn't take any more money out of his account, on which week will his balance reach £100?

   _____

8. Tick which is more likely:
   - seeing a man who is over 200 cm tall
   - seeing a man with a mass of more than 100 kg.

9. On the scale below, which letter best represents the likelihood of the spinner scoring: 4?

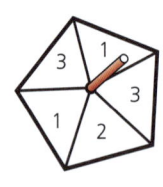

   _____

   A   B   C   D   E   F
   |----|----|----|----|----|
   Impossible   Even chance   Certain

10. On the scale above, which letter best represents the likelihood of picking a card with **R** from Barry's letter cards below?

    _____

    B A R R Y

Mixed test 1

Final score

20 ___ %

1 (a) On the number grid, shade all the multiples of 3 and circle all the multiples of 5

1	2	3	4	5	6
11	12	13	14	15	16
21	22	23	24	25	26
31	32	33	34	35	36

(b) Write down a number between 40 and 50 which is a multiple of both 3 and 5
_____

2 (a) Write **two thousand, one hundred and five** in figures. _____

(b) Write 3056 in words. _____
_____

(c) Write in figures the Roman number CLVII.
_____

3 (a) Round 1446 to the nearest 10
_____

(b) Round 97 056 to the nearest 100
_____

4 (a) Shade half of this shape.

(b) Complete these equivalent fractions:

$\frac{1}{3} = \frac{}{12}$

(c) Write these fractions in order of increasing size: $\frac{1}{2}, \frac{1}{8}, \frac{1}{3}$

_____ _____ _____

(d) Write $1\frac{1}{3}$ as an improper fraction.
_____

5 Complete these calculations:

(a) $2 \times 4 - 3 =$ _____

(b) $2 \times (4 - 3) =$ _____

6 Given that $11 \times 15 = 165$, complete the following:

(a) $11 \times 150 =$ _____

(b) $11 \times 30 =$ _____

(c) $1650 \div 15 =$ _____

7 (a) Add: $347 + 734$

_____

(b) Subtract: $110 - 73$

_____

8 The result of a calculation was 2.25 What would this mean in

(a) hours and minutes

_____ hours _____ minutes

(b) pounds and ounces?

_____ lb _____ oz

9 Complete these statements using **even** or **odd**:

(a) An odd number plus an _____ number gives an even number.

(b) An _____ number times an _____ number gives an odd number.

10 The nutrition information on a shortbread finger is:

	per 100 g
Fat	25 g
Carbohydrate	65 g
Protein	5 g

(a) What fraction of a biscuit is fat?
_____

(b) One finger contains 5 grams of fat. A pack of shortbread fingers has a mass of 300 g. How many fingers are in a pack?
_____

67

Mixed test 2

Final score

20 ___ %

Questions 1 to 2 concern the sequence of patterns.

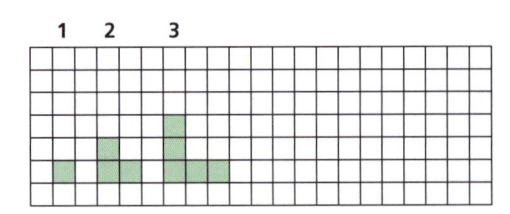

1 2 3

1 Draw pattern **4** and pattern **5** in the sequence.

2 Complete this table of data.

Pattern number	1	2	3	4	5
Area (square units)	1	3			
Perimeter (units)			12		

3 Find the number represented by the symbol in each of these sentences.

(a) 27 + 🔵 = 72   🔵 = _____

(b) ✽ − 23 = 32   ✽ = _____

4 In this sequence of beads, the 1st bead is red and the 2nd bead is purple.

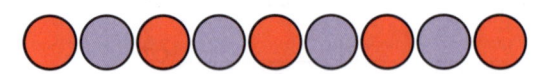

What will be the colour of the

(a) 15th bead

(b) 100th bead?

5 The machine below adds 1 to every number put in, giving pairs of (in, out) numbers.

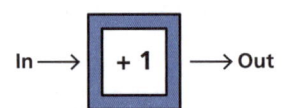

In → [+1] → Out

For example: 3 goes in, 4 comes out, (3, 4).

On the grid, plot 3 more (in, out) number pairs.

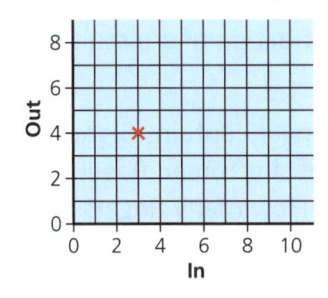

6 What are the readings on the scale?

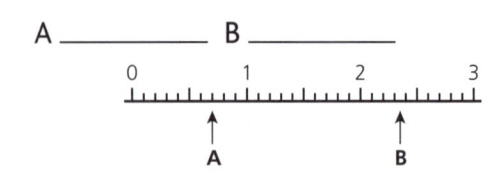

A _____   B _____

7 Complete the shape, which is symmetrical about the line **m**.

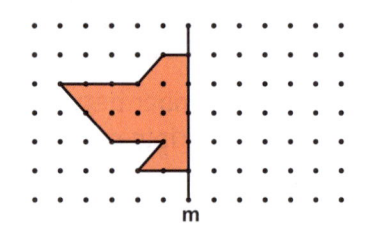

8

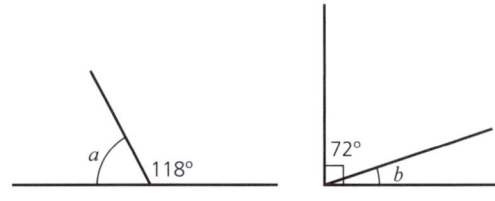

(a) Calculate angle a. _____°

(b) Calculate angle b. _____°

9 The bar chart shows the sales of ice-creams one day.

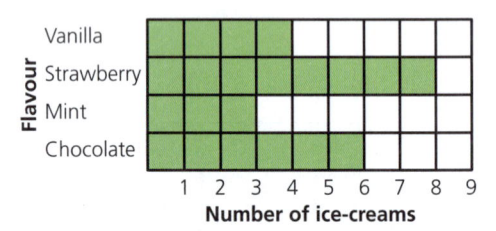

If the ice-creams cost £1.20 each, how much money was taken?

£_____

10 Tick which is more likely:

- scoring 3 or more when an ordinary die is rolled
- getting 'heads' when a coin is tossed.

68

Mixed test 3

Final score

20 ___ %

1 (a) What is the largest number which is a factor of both 12 and 30? _____

   (b) Complete the factor rainbow to show all of the factor pairs for 40

   1 2 ___ ___ ___ ___ ___ 40

   (c) Count backwards in 100s from 1120

   1120 ____ ____ ____ ____

2 (a) Write **two thousand and seven** in figures. _____

   (b) Write 305 019 in words.
   _____
   _____

   (c) What is the value of the 7 in 17 019? _____

3 (a) Round 51 495 to the nearest 100 _____

   (b) Round 11.51 to the nearest whole number. _____

   (c) Round 5.35 to 1 decimal place. _____

4 (a) Write the fraction $\frac{89}{100}$ as a decimal. _____

   (b) Write the decimal 0.8 as a fraction in its simplest form. _____

   (c) Write 30% as a fraction. _____

   (d) Write 28% as a decimal. _____

5 Complete the multiplication square.

×	9	6	8	7
6				
8				
7				
9				

6 (a) Subtract 599 from 1000 _____

   (b) How many books priced at £5.99 could you buy with a £20 note? _____

   (c) Find the total cost of 5 chocolate bars priced at 69p each. £_____

7 (a) Multiply: 235 × 12

   _____

   (b) Divide: 3012 ÷ 6

   _____

8 The result of a calculation was 5.4 What would this mean in

   (a) hours and minutes

   _____ hours _____ minutes

   (b) centimetres and millimetres?

   _____ cm _____ mm

9 Zoe multiplied 409 by 28

   (a) What is the units digit of the correct result? _____

   (b) By rounding both numbers, write down an approximate value for the result.

   _____

10 The diagram below shows information about the children in Year 5

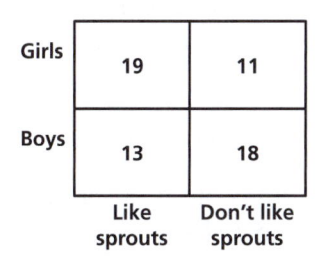

	Like sprouts	Don't like sprouts
Girls	19	11
Boys	13	18

   (a) How many children in Year 5 said that they do not like sprouts? _____

   (b) How many children are in Year 5?

   _____

69

Mixed test 4

Final score
20 ___ %

Questions 1 and 2 concern applying a rule over and over again to 2-digit numbers.

The rule is: Double the tens digit and then add the units digit. Stop when you reach a single-digit number. For example:

34 → 10 → 2   83 → 19 → 11 → 3

1. Show what happens when the rule is applied to:

    (a) 47 _____

    (b) 99 _____

2. Find a number between 60 and 70 that reaches the single-digit 9 _____

3. Find the number represented by the symbol in each of these sentences.

    (a) 23 = 36 − ☞   ☞ = _____

    (b) ✲ ÷ 5 = 15   ✲ = _____

4. In the sequence of letters below, the 3rd, 6th and 9th letters are all **A**.

    A B A A B A A B A A B

    What will be

    (a) the 18th letter _____

    (b) the 20th letter? _____

5. Sally has a machine, but the label has fallen off and she can't remember what it does!

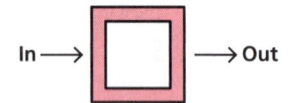

    In → ☐ → Out

    If Sally puts in 1, 3 comes out. If she puts in 2, 4 comes out.

    (a) Write the operation on the label.

    (b) If 12 comes out of Sally's machine, what number did she put in?

    _____

6. All the water is poured from **B** into **A**.

    Draw the new water level in **A**.

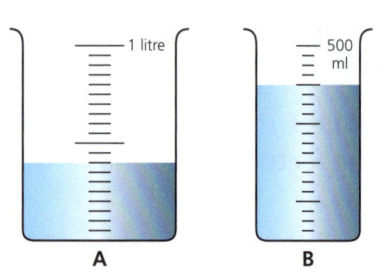

7. Complete the shape, which is symmetrical about the line **m**.

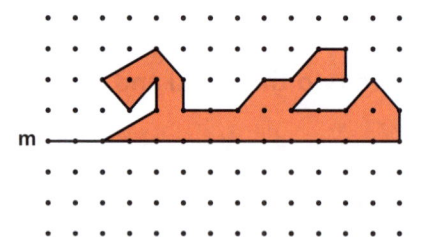

8. (a) On the co-ordinate grid, plot the points A (2, 2), B (5, 2), C (7, 7) and D (4, 7) and join them to make shape ABCD.

    (b) Name the shape formed by joining points A, B, C and D in order. _____

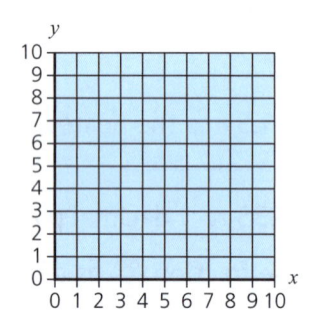

9. The table records data about four friends, all aged ten at Christmas.

Name	Birthday	Height (cm)	Mass (kg)
Simon	23 May	132	42
Lara	11 October	141	41
Una	29 February	138	44
Victor	14 June	142	41

(a) Who is youngest? _____

(b) Who is second tallest? _____

10. One of the friends is chosen at random. What is the likelihood of the friend having a mass of 41 kg?

_____

70

Mixed test 5

Final score
20 ___ %

1 (a) Circle the prime numbers in the list below.

13   23   33   43   53   63   73

(b) What temperature is 7 degrees higher than ⁻3 °C? _____ °C

(c) What is the square of 8? _____

2 (a) What is the value of the 4 in 340 000? _____

(b) Draw beads on the abacus to represent the number which is 5 more than 909 997

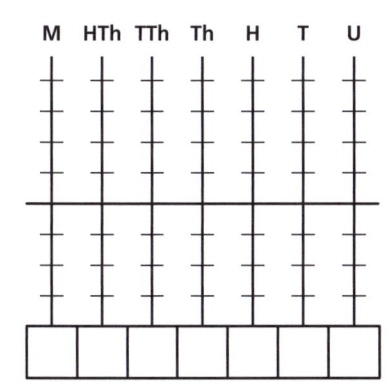

M   HTh   TTh   Th   H   T   U

(c) Multiply 308 by 10 _____

3 (a) On the scale, mark the positions of ⁻3 and 6.5

0   5   10

(b) Round 441 to the nearest 100 _____

(c) Round 9.65 to 1 decimal place. _____

4 (a) Shade $\frac{1}{4}$ of this shape.

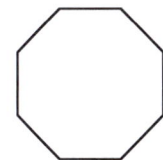

(b) Complete these equivalent fractions:

$\frac{3}{5} = \frac{}{20}$

(c) Write these fractions in order of increasing size: $\frac{3}{5}, \frac{3}{7}, \frac{3}{4}$

_____   _____   _____

5 (a) Write the other three addition and subtraction facts using the same numbers.

43 − 29 = 14   _____ − _____ = _____

_____ + _____ = _____

_____ + _____ = _____

(b) Write the other three multiplication and division facts using the same numbers.

156 ÷ 12 = 13   _____ ÷ _____ = _____

_____ × _____ = _____

_____ × _____ = _____

6 Given that 35 × 48 = 1680, complete the following:

(a) 35 × 24 = _____

(b) 36 × 48 = _____

7 (a) Divide, giving your answer with a remainder: 40 ÷ 7 _____

(b) Divide, using factors: 420 ÷ 28 _____

8 The result of a calculation was $2\frac{3}{4}$. What would this mean in

(a) hours and minutes

_____ hours _____ minutes

(b) feet and inches?

_____ feet _____ inches

9 Complete the statements using **even** or **odd**:

(a) An even number multiplied by an odd number gives an _____ number.

(b) An odd number minus an _____ number gives an odd number.

10 Lucy buys 6 cakes priced at 85p each and 6 cups of tea priced at £1.15 each.

(a) How much change will Lucy receive from a £20 note? £_____

(b) In change, Lucy receives a £5 note and 2 coins. What are the coins? _____

Mixed test 6

Final score
__/20 ___%

Questions 1 and 2 concern the sequence of patterns below.

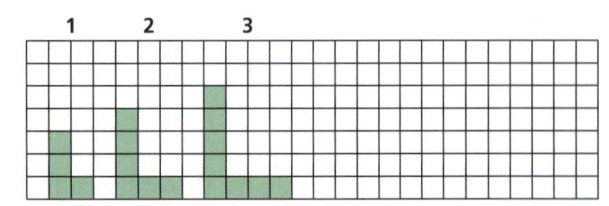

1 2 3

1 Draw pattern **4** in the sequence.

2 How many squares are there in pattern 6? _____

3 On each line write the operation sign (− + × ÷) to make the sentences true.

(a) 54 _____ 9 = 63

(b) 25 _____ 5 = 7 _____ 2

4 Anne is making up a string of 40 beads with the letters of her name.

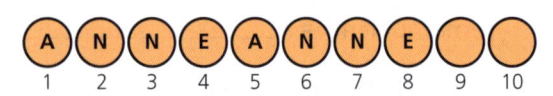

A N N E A N N E
1 2 3 4 5 6 7 8 9 10

(a) What letter will the last bead be? _____

(b) How many **N** beads will there be? _____

5 Aled has a machine which divides any number that he puts into it by 2

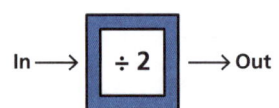

In → ÷ 2 → Out

If $5\frac{1}{2}$ comes out, what number did Aled put in? _____

6 The diagram shows a rectangle.

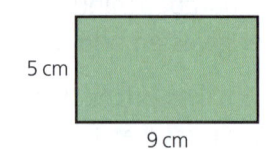

5 cm
9 cm

Calculate

(a) the area _____ cm²

(b) the perimeter. _____ cm

7 Name the plane shapes below.

A _____ B _____
C _____ D _____

A B C D

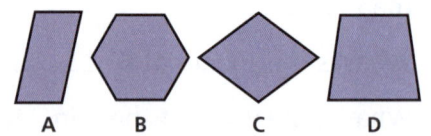

8 (a) Name the types of these angles.

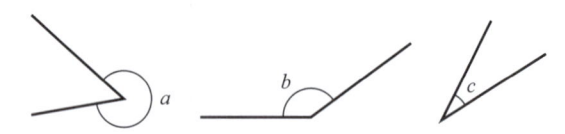

a _____ b _____
c _____

(b) Calculate angles d and e.
$d =$ _____ ° $e =$ _____ °

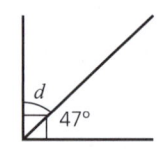

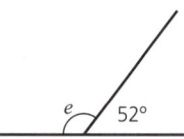

d 47°   e 52°

9 The tally below records the scores when a die is rolled.

Score	Tally of scores	Frequency
1	𝍩𝍩𝍩 III	18
2	𝍩𝍩 II	
3	𝍩𝍩𝍩 𝍩 I	
4	𝍩𝍩𝍩 II	
5	𝍩𝍩𝍩	
6	𝍩𝍩𝍩 II	

(a) Complete the table of frequencies.

(b) What percentage of the rolls resulted in a score of 4 or more? _____ %

10 Which letter on the scale best represents the likelihood of getting 'heads' when a coin is tossed? _____

A B C D E F G
Impossible   Even chance   Certain

Mixed test 7

Final score
20 ___ %

1 Write down

   (a) the square of 11 _____

   (b) 5³ (the cube of 5) _____

   (c) all of the multiples of 3 between 40 and 50 _____

2 (a) If these numbers are arranged in order of size, which will be in the middle?

   _____

   3476   3764   7346   6347   4763

   (b) Write down the number which is exactly half way between 43 and 71

   _____

3 (a) On the scale, mark the position of 4

   0 _____ 10

   (b) Round 379 599 to the nearest 1000

   _____

   (c) Round 8.95 to 1 decimal place.

   _____

4 (a) Add: $\frac{1}{5} + \frac{3}{5}$

   _____

   (b) Subtract: $\frac{5}{8} - \frac{3}{8}$

   _____

   (c) Multiply: $\frac{2}{5} \times 100$

   _____

5 Complete the multiplication square.

×	5	9	7	6
7				
9				
8				
6				

6 (a) What is the sum of the numbers from 20 to 24 inclusive?

   _____

   (b) Subtract 504 from 1000 _____

7 (a) Multiply: 47 × 23

   _____

   (b) Divide: 5040 ÷ 9

   _____

8 Picture framing comes only in 2.8 m lengths. Vaibav calculated that he needed about 0.8 metre for each picture he wanted to frame. He divided 2.8 by 0.8 and the result was 3.5

   (a) How many pictures can he frame from one length of framing?

   _____

   (b) What length of framing will be left over?

   _____ cm

9 Philippa multiplied 795 by 22

   (a) What is the units digit of the correct result?

   _____

   (b) By rounding both numbers, write down an approximate value for the result.

   _____

10 The timetable shows times for 2 trains A and B. Both trains take the same time between the towns. Complete the timetable.

	A	B
Peely	07:59	11:15
Quewty	09:30	___ : ___
Ruffly	___ : ___	15:08

73

Mixed test 8

Final score

20 ____ %

Questions 1 and 2 concern applying a rule over and over again to 2-digit numbers.

The rule is: Multiply the units digit by two and then add the tens digit. Stop when you reach a single-digit number. For example:

18 → 17 → 15 → 11 → 3   53 → 11 → 3

1 Show what happens when the rule is applied to

(a) 39 _____

(b) 98 _____

(c) 49 _____

2 Find four numbers between 10 and 20 that give the single-digit result 3 _____

3 Find the number represented by the symbol in each of these sentences.

(a) ✳ + 29 = 53   ✳ = _____

(b) 44 − ❖ = ❖   ❖ = _____

4 Tina is making a necklace of beads with the letters of her name

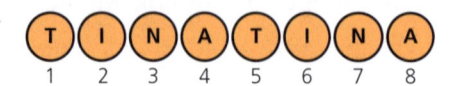

T I N A T I N A
1 2 3 4 5 6 7 8

The beads in positions 1 and 5 are **T**.

What will the 19th bead be? _____

5 The machine below subtracts 2 from any number put into it.

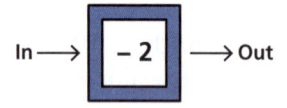

In → −2 → Out

If 3 is put in, 1 comes out. (3, 1) On the grid, mark with a cross the point representing the (in, out) number (3, 1) and plot 3 more (in, out) points for this machine.

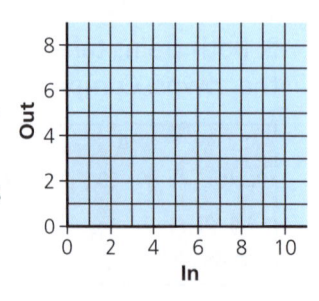

6 Write down the readings on scales **A** & **B**.

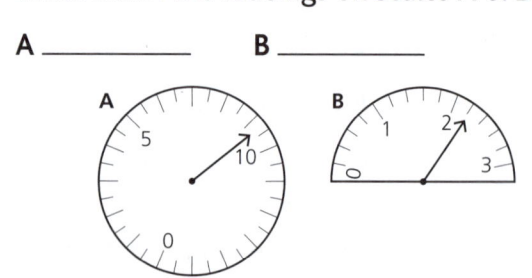

A _____   B _____

7 Draw all the lines of symmetry on the plane shapes below.

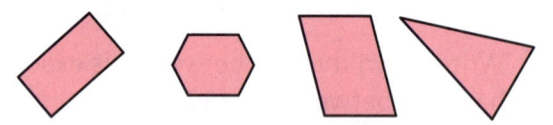

8 (a) On the co-ordinate grid, plot the points A (2, 6), B (5, 8), C (8, 6) and D (5, 4). Join the points in order to make shape ABCD.

(b) Translate ABCD 4 units down.

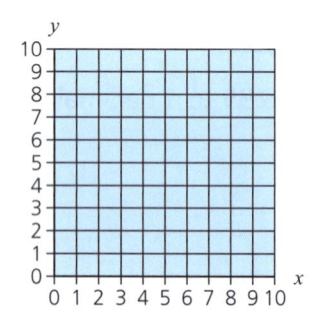

9 The table records data about four friends.

Name	Age (y : m)	Height (m)	Mass (kg)
Kiera	9 : 10	1.37	39
Lara	9 : 8	1.40	41
Mia	9 : 11	1.41	43
Niamh	10 : 3	1.39	40

(a) Who was born first? _____

(b) Who has the smallest mass? _____

10 If a friend is chosen at random, what is the likelihood of her being at least 1.35 m tall?

_____

Mixed test 9

Final score
20 ___ %

1. What is the smallest 3-digit common multiple of 3 and 7? _____

2. (a) Write in figures the Roman number MMDCCCLXXVIII. _____

   (b) If these numbers are arranged in order of increasing size, which will be in the middle? _____

   43.5   4.35   4.53   3.45   35.4

   (c) Write down the number which is half way between 56 and 102 _____

3. (a) Round 205 950 to the nearest 100
   _____

   (b) Round 2 499 999 to the nearest million.
   _____

   (c) Round 13.6 to the nearest whole number.
   _____

   (d) Round 17.99 to 1 decimal place.
   _____

4. (a) Complete these equivalent fractions:

   $\frac{8}{12} = \frac{}{3}$

   (b) Add: $\frac{5}{12} + \frac{11}{12}$
   _____

   (c) Multiply: $\frac{3}{4} \times 48$
   _____

5. Complete the multiplication square.

×	6	12	7	9
8				
7				
9				
11				

6. (a) What is the sum of the first four square numbers?
   _____

   (b) Subtract £11.89 from £20.00 £_____

7. (a) Multiply: 317 × 12

   _____

   (b) Divide: 1001 ÷ 7

   _____

8. The result of a calculation was 2.05 What would this mean in

   (a) hours and minutes

   _____ hours _____ minutes

   (b) litres and millilitres?

   _____ litres _____ ml

9. Complete, using **even** or **odd**:

   (a) An even number multiplied by an even number gives an _____ number.

   (b) If an odd number divides exactly by an odd number the result is an _____ number.

10. The nutrition information on a snack bar is:

	per 100 g
Fat	5 g
Carbohydrate–sugar	50 g
Carbohydrate–other	25 g
Protein	4 g

    What fraction of the carbohydrate is sugar?
    _____

75

Mixed test 10

Final score
20 ___ %

Questions 1 and 2 concern the sequence of patterns below.

1 Complete this table of data.

Pattern number	1	2	3	4	5
Number of small triangles		4			
Perimeter (units)	3		9		

2 For pattern 8, what would be the number of small triangles?

_____

3 On each line write the operation sign (− + × ÷) to make the sentences true.

(a) 24 _____ 8 = 3

(b) 24 _____ 8 = 2 _____ 8

4 Sara is making up a string of 60 beads with the letters of her name.

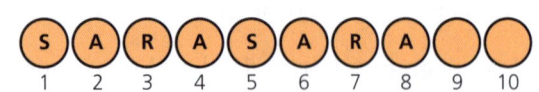

S A R A S A R A
1 2 3 4 5 6 7 8 9 10

(a) What letter will the last bead be? _____

(b) How many **A** beads will there be? _____

5 Claire has a machine which multiplies any number that she puts into it by 3

In → × 3 → Out

If $1\frac{1}{2}$ comes out, what number did Claire put in?

_____

6 The diagram shows a rectangle.

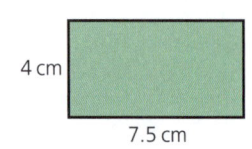

4 cm
7.5 cm

Calculate

(a) the area _____ cm²

(b) the perimeter. _____ cm

7 Name the plane shapes below.

A _____ B _____

C _____ D _____

A B C D

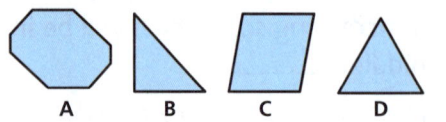

8 (a) Name the types of these angles.

a _____ b _____

c _____

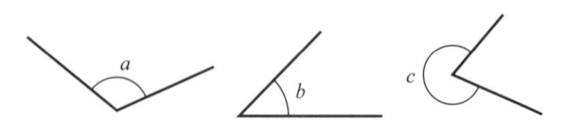

(b) Calculate angles d and e.

$d =$ _____° $e =$ _____°

49°   116°

9 In the pictogram below, one symbol represents **4** butterflies.

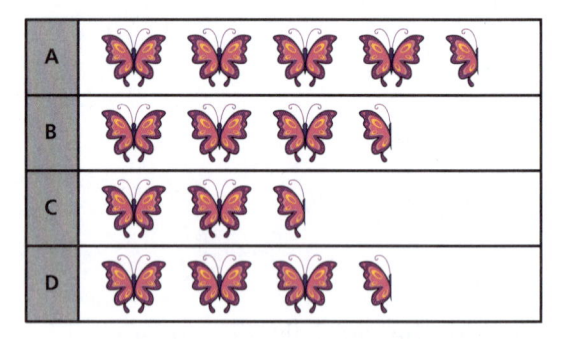

A	🦋 🦋 🦋 🦋 🦋
B	🦋 🦋 🦋 🦋
C	🦋 🦋 🦋
D	🦋 🦋 🦋 🦋

What is the total number of butterflies?

_____

10 Which letter on the scale represents the likelihood of getting a cube number when a die is rolled? _____

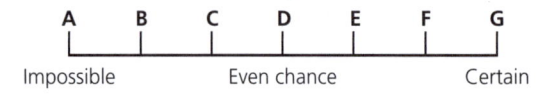

A B C D E F G
Impossible   Even chance   Certain